2030

2030

TECHNOLOGY THAT WILL CHANGE THE WORLD

Rutger van Santen
Djan Khoe
Bram Vermeer

OXFORD
UNIVERSITY PRESS

2010

OXFORD
UNIVERSITY PRESS

Oxford University Press, Inc., publishes works that further
Oxford University's objective of excellence
in research, scholarship, and education.

Oxford   New York
Auckland   Cape Town   Dar es Salaam   Hong Kong   Karachi
Kuala Lumpur   Madrid   Melbourne   Mexico City   Nairobi
New Delhi   Shanghai   Taipei   Toronto

With offices in
Argentina   Austria   Brazil   Chile   Czech Republic   France   Greece
Guatemala   Hungary   Italy   Japan   Poland   Portugal   Singapore
South Korea   Switzerland   Thailand   Turkey   Ukraine   Vietnam

Published by Oxford University Press, Inc.
198 Madison Avenue, New York, New York 10016

www.oup.com

Oxford is a registered trademark of Oxford University Press

Library of Congress Cataloging-in-Publication Data
Santen, R. A. van (Rutger A.)
2030 : technology that will change the world / Rutger van Santen,
Djan Khoe, Bram Vermeer.
p.   cm.
Includes bibliographical references and index.
ISBN 978-0-19-537717-0
1. Technological forecasting.
I. Khoe, Djan. II. Vermeer, Bram. III. Title.
T174.S263 2010
601'.12—dc22   2009053883

1 3 5 7 9 8 6 4 2

Printed in the United States of America
on acid-free paper

CONTENTS

2030

Part 0
BASICS

0.0

OUR MISSION

There was no shortage of unprecedented events as we were writing this book. Oil and food prices rocketed and then fell back to Earth; there was a devastating earthquake in Haiti; banks failed; and a new flu virus sparked a worldwide pandemic alert. None of these developments was predicted a year in advance—or at least, not loudly enough to be heard. For all our technological and forecasting skills, we proved unable to take appropriate measures in advance.

Technology has been helping us satisfy our material needs since prehistoric times. We learned how to till the soil, how to communicate with one another, and how to stay healthy. Almost everyone in the Western world now has enough to eat, a roof over their heads, and clean water. A great many basic needs have therefore been met—so much so that some observers now claim that the need for further technological advances is diminishing. Recent events argue against such a view. Humanity is increasingly confronted with crises that, for the first time in our history, are global in scope. The food shortages we saw in 2007 occurred simultaneously in Asia, Africa, and South America; the recession that took hold in 2008 did so simultaneously worldwide; and when the flu pandemic broke out in 2009, germs were able to cross between continents in a matter of days. Climate change and oil depletion, meanwhile, are no less global challenges that we will face in the decades ahead. The globalization of disaster is itself rooted in our technology. Generations of engineers have steadily woven an international web of industries, communications, and markets that has resulted in planetary interdependence. These global networks are now so tightly knit that we share a common fate. We will now survive together or quite possibly perish together.

The authors of this book are concerned about the new scale on which many of these pressing problems are now manifesting themselves. Because technology has been a key factor in triggering these issues in the first place, we believe it should also be part of solving them and of preventing similar problems from arising in the future. The scale of the challenges is different from anything that we humans have so far experienced, and so the solutions need to be different, too—even in the case of problems we thought we had already beaten. We might know how to cure a virus infection, for instance, but curing billions of people simultaneously is a very different challenge and one that will require new technology.

We clearly couldn't cover every field of technology ourselves, so we talked to numerous scientific and technological experts and visionaries about how they'd like the world to be in 20 years. We asked them to explore the kind of research that will be necessary in the years ahead and to put it in its broader context. Do we have the means to influence the course of history? What breakthroughs will be needed to make the world a better place? Despite their very different areas of research, the experts presented ideas that displayed a number of important parallels. They showed us that many processes have become interrelated and that global networks of all kinds are now intertwined. This means that a disturbance in, say, the Internet now has the potential to disrupt the global financial system. Science and technology are closely implicated in the growing complexity of key issues—something that complicates our view of both the problems and their solutions.

In the course of our discussions with the experts, we gradually realized, therefore, that the relevance of networks and complexity is much wider than we thought, providing us with a new way of viewing problems and solutions. Although this has yet to be fully appreciated in many branches of science, interest in the approach we describe here is definitely increasing. It also presents us with new tools with which to face the future. Complex processes lack regularity and predictability, so we certainly can't predict the future. Nevertheless, complexity science has a great deal to teach us about breakthroughs, changes, and patterns of influence. Research into complex dynamical systems has grown into a new science since the late 1980s. It is now common in physics, chemistry, and mathematics and is gradually finding its way into other disciplines as well. Fresh understanding of the regularities underlying complex systems offers us a new perspective on sustainability, stability, and crisis-prevention methods—a new way of looking forward that can help us identify key issues for 2030.

The 20-year horizon was chosen carefully. For many scientists, 2030 doesn't feel like the distant future. We don't have to fantasize about the solutions that might be available, as many of the technologies we will be using around that date are a logical extension of what we're already seeing in our labs today. Two decades is "doable" as a period for scientists to envision. Technical developments are often long in the making, and a lot of the ideas being developed today will need that kind of timescale to become reality. So you won't find any science fiction stories in this book about jetpacks or robots taking over the world. Maybe those things will happen one day, but probably not in the next 20 years.

On the other hand, two decades is far enough ahead to take us beyond our immediate needs. This book is not about incremental improvements to existing technologies or the next generation of microchips. It does not offer projections of the population 20 years from now, how many cars we'll own, or how many hospital beds we'll need. We're not interested in statistics or in developing scenarios but in the major bottlenecks our society will have to deal with, and we don't need detailed numbers to demonstrate what those are. It was for all of these reasons, then, that we asked the experts featured in this book to focus on the key issues of 2030.

As technologists, we feel a special responsibility for providing the means to work toward solutions. Technology has been a major contributor to many of the problems we discuss here, so it is also our duty to help solve them. Like so many of our colleagues, we have placed a great deal of our creativity at the service of industry. This is important, but it does not sufficiently solve the major problems humanity now faces. There are many themes that should urgently be addressed by anyone who wishes to make the world a better place. And that definitely includes technologists like us. In this book, we show that technology has the power to make a genuine difference. We need a variety of technological breakthroughs to help secure our future. At the same time, we should carefully monitor the relationship between the different disciplines. Ideally, this book will make its own modest contribution to the research agenda by inspiring engineers and others in their efforts to protect and improve our planet.

This international publication grew out of a survey of future developments we produced to mark the fiftieth anniversary of Eindhoven University of Technology in the Netherlands.[1] The encouraging response we received from our fellow engineers in the Netherlands and outside and the global nature of the issues now confronting our planet encouraged us to expand our study on a more global scale.

0.1

CONCERNS

When we asked our colleagues to list what they consider the most pressing problems facing our planet today, they came back with a wide range of concerns, including atmospheric pollution, climate change, intensifying security threats, and the need to secure an adequate food supply for all the world's people. Given that we concentrate in this book on the most crucial of these issues, it is dispiriting that we should still have to begin with the most basic of human needs. In the early twenty-first century, a lack of food, water, and shelter continues to rob tens of millions of people of their lives every year. More than half of all the world's deaths are attributable to malnutrition.[1] More people die of hunger every year than perished in the whole of World War II. What makes this problem especially distressing is that we know it isn't necessary. That's why we give particular prominence in this book to the question of what we should do about it.

Once the basic necessities have been taken care of, the next biggest killers are cancer and infectious diseases. And as we grow older, we become increasingly concerned about our reliance on caregivers and the decline in our cognition. Breakthroughs in these fields would enable us to live longer and happier. So that's the second category of challenges we discuss in this book. The continued existence of the human race is not, of course, guaranteed. The rapid pace of change on our planet requires us to adapt significantly and quickly. We discuss the issues arising from that recognition in a separate part of the book devoted to the sustainability of our Earth. The stability of our society isn't guaranteed either. Financial crises, explosive urban growth, and armed conflict all have a detrimental effect on our well-being, making this the fourth category of the problems we consider.

In our view, the most important issues human beings need to work on are: malnutrition, drought, cancer, infectious diseases, care of the elderly, cognitive deterioration, climate change, depletion of natural resources, natural disasters, educational deprivation, habitable cities, financial instability, war and terrorism, and the infringement on personal integrity.

These problems are closely interrelated at a deeper level, as we describe in this book. We do not, therefore, attempt to rank them. Anyone working to solve the issue of climate change is helping to combat poverty as well. The intertwining of these problems also means that we can develop a common set of tools—the subject of a central chapter in this book. Understanding of communication techniques, computers, and logistics can be helpful on many different fronts.

We are by no means the only people concerned about the future. In 2004, Danish environmental activist Bjørn Lomborg organized a conference in Copenhagen to discuss the major issues currently facing the global community.[2] It closely resembled what we are trying to describe here. Lomborg, who has come under intense fire from environmentalists and mainstream scientists alike, was keen to draw up an accurate balance sheet of the current state of the planet. The experts who took part produced a list of the most important challenges, the top ten of which were labeled the "Copenhagen Consensus." Much like our own list, it includes climate change, infectious diseases, armed conflict, education, financial instability, government and corruption, malnutrition and starvation, population and migration, hygiene and water, subsidies, and trade barriers. The outgoing president of the American Association for the Advancement of Science (AAAS), meanwhile, presented another list in 2007,[3] which again reflects the UN's Millennium Development Goals.[4] There is a broad consensus, then, regarding the principal challenges we face.

For all our technical advances, the list of crucial issues facing humanity has barely changed over the past century. British science-fiction writer H. G. Wells was one of the first to publish futurological, nonfiction essays around 1900.[5] Wells was convinced that the twentieth century would pose enormous threats to the human race. The nineteenth century, he argued, had unleashed an unstoppable wave of progress that would radically disrupt people's lives, generate social unrest, and trigger wars. Wells believed that new means of transportation in particular would change the face of the planet. The victory of roads over railways would devastate cities, and those same cities—having themselves been shaped by the railway—would now proliferate across the countryside and generate an endless sprawl of suburban settlements. Wells predicted that military engineers would design some kind of "land battleship," and "very probably before 1950, a successful aeroplane will have soared and come home safe and sound." He foresaw terrible aerial battles but also the kind of changes that would likely take place in the household. Working in the kitchen, he wrote, would cease to be a full-time occupation. He arguably predicted globalization, too, raising issues that overlap with those we discuss here.

The themes raised by Wells were reiterated throughout the twentieth century, often using different terminology that reflected the anxieties of the day. We are convinced that there is nothing inherent in human nature that prevents us from solving these problems. To do so, however, we first need to understand the mechanisms that perpetuate them. Only then might we hope to work toward a successful remedy. We believe that solutions are needed that can be applied and broadly supported at a global level. It no longer makes sense to combat disease on a local scale. More disciplined use of antibiotics in the human population is pointless, for instance, if the same discipline is lacking in the treatment of animals. In the next chapter, we explore the common mechanisms that underlie many of the issues we face today.

0.2

APPROACH

It might seem a little foolish to attempt to predict the future, especially when you consider how often earlier predictions failed. For instance, the imminent depletion of our main fossil fuel reserves has often been proclaimed. Back in 1865 Stanley Jevons predicted that Britain's coal reserves would run out within a few years. The U.S. government calculated in 1914 that there was enough oil left in the ground for just one more decade; subsequent forecasts in 1939 and again in 1951 predicted that the oil would be gone within 13 years. In the 1960s, optimism about the advent of nuclear energy was so high that it was actually argued that the available gas and oil should be consumed as quickly as possible because the advent of virtually cost-free nuclear energy would soon render them worthless. In 1972, the Club of Rome was forecasting that we had only enough oil for another 20 years.[1] *Peak oil*—the moment at which consumption reaches its historic plateau—was called during the 2008 surge in oil prices, only for the prophets of doom to clam up when prices began to fall again. Nowadays, if you ask different energy experts how long our oil reserves will last, their projections will vary by a factor of at least three, and some will simply reply, "forever."

There are many more examples we could add. Predictions for future events fail time and again. The reason should be obvious when you consider major developments in our past. Small haphazard events can change the course of history. One spark of genius by a German mechanic changed the world of transport forever. A single decision by an Arab sheikh could bring our oil-based economy to its knees. Or a new battery technology could change our approach to transportation forever. Turning points like these can't be foreseen. They depend on one person's thought processes or on a stroke of experimental luck by a group of scientists. History can only be recounted in hindsight. That means futurologists have their work cut out for them: They can merely extrapolate from trends that are already visible. Sometimes, they hedge their bets by presenting several different scenarios that vary according to the precise economic and social circumstances. This is a very useful approach in preparing for short-term, gradual changes and for learning to think out of the box.

But often, the scenarios are so divergent that they offer no tool for longer-term planning. The problem is that those scenarios don't present really new information, as it is an extrapolation from a known situation. Gradual extrapolations miss sudden changes. Stability can be fragile and misleading. Another problem is that those scenarios only give numbers and statistics; they're about phenomena rather than forces that drive them. Scenarios often extrapolate only certain aspects of societal development, leaving other aspects unchanged. Especially, future scenarios often use a "current level of technology," thereby missing the drive of engineers to solve really pressing problems. Or worse, they assume that there is nothing left to be discovered (such as in many peak oil predictions). Useful as those methods may be for strategic thinking, they don't give us an understanding, and they are sure to miss important turning points in our future.

For setting an agenda for action, a new approach is being developed to supplement the toolboxes of futurists. In the past decade, physicists, chemists, biologists, and sociologists have worked side by side to unravel the patterns underlying complex phenomena like earthquakes, biological evolution, and ethnic violence. They have laid the foundations of the new science of collective phenomena in society and the natural environment. And frequently to their own surprise, they have identified regularities that help predict change and stability. There is a greater amount of order beneath the seemingly haphazard facts of history than we often think. This research has given us the tools to identify turning points and also inspired policies with the potential to steer us away from catastrophe.

THE NEW SCIENCE OF COLLECTIVE PHENOMENA

To understand this fresh approach, it is important to realize how the new science differs from the methods scientists have used traditionally. When Galileo Galilei supposedly dropped balls of varying size from the Leaning Tower of Pisa, he noted that they all fell at the same rate. Or rather they didn't because air resistance interferes with the process. Nevertheless Galileo carefully designed experiments that rendered the influence of the surrounding air negligible, leaving him with only the gravitational force. After so many years, it's not at all clear which experiments Galileo really did, but the story is recounted over and again because it symbolizes the way science has since worked. Concentrating on a single aspect of reality in this way often proved to be a very powerful technique.

Scientists have continued to focus on isolated phenomena like this for centuries. Regularity is most obvious when we study a process that has been separated from its surroundings, enabling us to concentrate entirely on a cause and its effect. This reductionist approach has enabled scientists to identify the primary laws of nature. The time it takes a pendulum to swing is, at first sight, directly proportional to the square root of its length. Similar relations exist for Ohm's law of electrical resistance and Newton's law as applied to the force acting on an apple. Or to phrase it the way mathematicians do when discussing proportionality: All primary laws have "strong linear terms."

Reductionist science has proved extremely successful. It has helped us understand, predict, and control nature. We can accurately forecast solar eclipses, for instance, well into the third millennium, in what is a formidable triumph of science over primitive speculation.[2] Reductionism has also been a powerful force beyond the realm of physics. Biologists derived simple laws for genetics by studying traits in isolation. And by the late eighteenth century, Adam Smith drew inspiration from the reductionist approach of contemporary physicists to formulate his fundamental economic laws. Many people at the time were surprised to discover that social phenomena like commerce could be described in terms of rules similar to those used in physics. Yet every trader shares the motivation of maximizing his or her profits, and it is that universality which creates the scope for mathematical abstractions.

In practice, the universal laws of reductionist science are mostly an oversimplification. Many phenomena don't occur in isolation. All manner of processes are frequently at work simultaneously, any of which can neutralize or intensify one another. As a result, there is no simple relationship in many cases between cause and effect. In mathematical terms, the descriptions of these processes often feature strong nonlinear terms. In 1961, American meteorologist Edward

Lorenz was surprised by the strange weather predictions generated by his computer. He discovered that a slight change in the starting conditions of his calculations could lead to a totally different forecast. Even rounding off his numbers in a different way was enough to produce a remarkable shift. Lorenz realized that this was not an artifact of his calculations but that the same occurs in real life—hence, his famous comment that the flap of a butterfly's wings has the potential to alter the course of the weather. Thus, minor perturbations can have a large effect in complex systems. Lorenz postulated that this makes the weather unforecastable more than a week in advance; its sensitivity to small changes renders the atmosphere "chaotic," as it would later be termed.[3]

The "butterfly effect" and "chaos theory" attracted a great deal of attention in both the scientific community and the media. The notion that the world around us seems to be developing at random evidently holds considerable appeal. Chaos theory was not, however, the end of the story. On the contrary, it merely marked the beginning of a scientific quest to trace regularity in unpredictable situations. It was found, for example, that most chaotic systems don't evolve completely at random. In many cases, they develop to a point at which some kind of stability is achieved. There may be more than one stable outcome, and evolution toward it may be entirely unpredictable, yet there is definitely order beneath the chaos. A simple example is the traffic on an expressway. When the number of vehicles is high, one of two possible situations will arise: Either everyone keeps moving at high speed, or a traffic jam will form. The line dividing the two is very thin. A small difference can cause a system to flip from one stable state into another. Traffic is liable to grind to a halt for no apparent reason. What's more, fluctuations often begin to appear as the critical boundary is approached, with the result that cars alternate between moving and stopping. Identifying those different end states (or "attractors") is one of the great achievements of chaos theory. Studying the dividing line between them could offer clues as to how we might push the situation in the right direction. Better expressway lighting, for instance, allows drivers to travel at higher speeds and respond to one another in a more controlled way. That means in turn that the traffic will be less likely to stop. In other words, the dividing line between the two attractors can be shifted. The example of expressway traffic may seem a little trivial, but many similar situations exist. The atmosphere above India, for instance, appears to have two states: monsoon and drought. Understanding the transition between the two could help improve agriculture in the region.

In the wake of Lorenz's groundbreaking results, a great many talented scientists set to work patiently unraveling patterns in complex dynamical systems. In the late 1980s, Danish physicist Per Bak began to study sudden transitions in complex systems—an area in which physics provides ample inspiration. Bak had used new insight from physics on how insulators can abruptly become conductors and observed how undercooled water can suddenly freeze. He then started to extend those same notions beyond the boundaries of physics. Bak confesses to a thick skin, which he believes was very helpful as he began to stick his nose into other people's disciplines. Exploring these new paths, he noticed that not every complex system moves toward stability. Some apparently evolve toward ever increasing *instability* instead. Tensions gradually build up in the earth's crust, generating greater and greater instability until a point is reached where they suddenly release their energy in an earthquake. Similar processes are at work in bodies of snow that progressively increase their mass until finally sliding down a mountainside as an avalanche. Other examples include forest fires and even mass extinctions of species.

These cataclysmic events are often repetitive. After an earthquake has occurred, the tension starts building all over again, and the cycle repeats. As long as the underlying forces remain, the earth's crust will move toward fresh instability and a new critical situation—something Per Bak terms "self-organized criticality." The next earthquake is certain to come; the only questions are when and how powerfully it will strike. If the tension is sufficiently high, a small random movement in the crust can trigger an earthquake. The details of each catastrophe might differ, but the forces at work are the same. Bak has studied the statistics of repetitive catastrophes. It was already known that the magnitude of successive earthquakes displays an odd regularity. An earthquake measuring eight on the Richter scale is quite rare. A magnitude seven quake, by contrast, is ten times more likely over any given period. Similarly, there will be around 100 tremors of magnitude six, 1,000 of magnitude five, and so on. This regularity is displayed over a very long period and also applies to different regions.[4] The proportion of smaller earthquakes to larger ones is therefore fixed, representing a "scaling law" or "power law" as mathematicians often call it. Scaling laws can be found where there may be a buildup of tension to a very high degree. They are characteristic of epidemics, wars, and even the growth of cities, stock market crashes, and famines, as we will see throughout this book.[5] Whenever we discover a scaling law, it's an indication that we need to study the underlying forces driving a situation out of equilibrium. A better understanding of those forces might help us interrupt the cycle of recurrent breakdowns and maybe prevent the next catastrophe.

Around the turn of the millennium, young scientist Albert-László Barabási pioneered a different approach to studying the criticality of complex systems.[6] Born an ethnic Hungarian in communist Romania, he began to study chaos theory just as Nicolae Ceauşescu was fighting to maintain his grip on power. Barabási realized that many complex situations can be interpreted as networks. Examples include the way a virus spreads through a web of relationships and the nature of an ecosystem as a network of predators and prey. This wasn't a new insight, but scientists had hitherto always used static networks to describe situations of this kind. What was new about Barabási's approach was his focus on how these networks change.[7] He discovered that the way they evolve reveals certain general patterns. This is because many networks regroup themselves to reach greater efficiency. It is often an advantage to connect to an already privileged link. This is why rich nodes in a network become richer. Commercial networks, say, tend to evolve until a number of privileged traders have taken possession of the pivotal points. Barabási also found that the distribution of wealth evolves until it comes to resemble a scaling law. A similar pattern of evolution turns out to be very common in different kinds of networks. The Internet, for instance, is governed by only a few multiconnected hubs, and the regulatory mechanisms of living cells have a few proteins that keep an entire range of processes in balance. Key connections like this are crucial to stability; if you take them away, the network fragments. The extinction of a few pivotal species could therefore cause a given ecosystem to collapse, while the survival of thousands of other species proves irrelevant. This knowledge could provide important clues in terms of conserving nature and in many other instances where networks feature a number of critical connections.

FUTURE PATTERNS

These are just a few of the fresh insights emerging from the new science of complex dynamical systems. The mathematics of complex problems has been further refined over the past decade with a wide range of new methods that enables us to pinpoint regularities. One of the important catalysts in the past decade was the Santa Fe Institute (U.S.A.) that created a research community for complexity themes that arise in natural, artificial, and social systems. Complexity has become a science in itself, and its methods are now firmly incorporated in the natural sciences and numerous fields of technology. They are used routinely when designing power, computer, and telecommunication networks and in advanced aircraft. At the same time, complexity science is beginning to filter through into the social sciences as well in areas like finance, economics, medicine, epidemiology, military conflict, and urban development.

Aspects of complexity science were touched on frequently as we conducted the interviews with experts for this book. Science and technology are being held back by the growing complexity of certain key issues, and so the methods of complexity science will be indispensable to any analysis of the challenges that lie before us. The last decade saw many interesting new publications by scientists who tried to apply the insights in complex systems to future developments of our society. In the field of climate research, for instance, it has become mainstream science to analyze future changes in terms of tipping points and transitions.[8] An influential essay by Thomas Homer Dixon speaks about the "thermodynamics of empire," which lead to tightly coupled socioeconomic and technological systems that may show sudden transitions.[9] Some think tanks and futurists have specialized in this type of analysis.

Stability and transition feature centrally in all scientific thinking about complexity. That's the first place we have to look when evaluating the common challenges facing us. Transitions can be repetitive as in the case of Per Bak's self-organizing critical systems. Or they might entail a flip from one stable situation to another as we find in a traffic jam. A process leading inexorably toward critical change might be going on unseen and unremarked below the surface. To give another example, it took almost a century from the beginning of the Industrial Revolution for anyone to notice that the burning of fossil fuels was having an impact on our climate.[10] Yet the signals were already there, making it vital that we learn to identify similar indicators at an earlier stage. A better and earlier understanding of sudden breakdowns like this might help us prevent similar catastrophes or, if that isn't possible, to delay such transitions. At least then we'd have the chance to prepare, making the change less painful than would otherwise be the case.

MODELING COMPLEXITY

Complexity science has benefited from advances in computer technology. Computers are ideal for mapping the simultaneous behavior of a multitude of processes, thereby helping us identify the key patterns of complexity. Computer calculations can't always lead to precise forecasts, however. Complexity scientists have learned the limitations of prediction from Edward Lorenz. Nonetheless, computer models can help us understand the underlying forces, interactions, and nonlinearities that together constitute the problem. Physicists can already calculate the collective behavior of a small crystal, taking account of each separate atom. They simultaneously track a few thousand atoms on their computers, which have a set of rules governing interactions among them. This enables scientists to calculate time step by time step how the collective is evolving. Together, the atoms constitute only a tiny crystal, yet the procedure gives us an accurate idea of how the properties of individual atoms can result in macroscopic phenomena. Similar calculations can be performed on interactions among immune cells, citizens, or businesses. Applied to a city, the principle resembles that of the popular computer game SimCity. In this instance, however, the game is deadly serious. To discover the forces driving the growth of a city, the rules are varied and the outcomes are compared with real-world situations, highlighting which forces are important and how they relate to one another.

As computing power increases, we can perform increasingly realistic "multiagent" or "Monte Carlo simulations," as physicists call them. Although many complex systems remain that are beyond the capacity of even the largest supercomputers, these too can be modeled if you apply a few mathematical tricks. The approach in this case is to perform detailed calculations for the crucial elements while making do with more coarse-grained techniques for more distant or slowly varying parts of the system. The challenge then is to "glue" the different levels of detail together. "Multiscale" strategies of this kind have been refined over the past decade in both physics and chemistry, where they link the microscopic scale of atoms with the macroscopic behavior of solids, fluids, and gases. The multiscale approach is increasingly used for other complex systems as well. Networks are especially suited to this approach, consisting as they often do of a mixture of remote and neighboring links that can be treated differently in a multiscale model.

WHAT CAN TECHNOLOGY DO IN THE FACE OF A CRISIS?

In this book, we identify many sudden transitions that we'd like to avoid. Technology often features prominently in the crises we describe. Our use of technology caused the greenhouse effect and has ensured that financial crises now occur on a truly global basis. As technology is part of the cause, it should also be part of the remedy, which is why it is important to focus on technology we can use to prevent catastrophes and to manage transitions. Technology might also help guide us away from certain critical transitions.

The first thing technology can do to help is to improve the accuracy of our measurements. Even if we can't predict the precise amount by which sea levels, say, will rise, we will at least know the direction in which we're headed. It's a matter of keeping your finger on the pulse, doing a lot of measurement, and using the results to check the outcomes suggested by computer models. A close-knit network of monitoring stations can tell us how ice sheets are changing and how rainforests are responding to climate change. Even earthquakes don't occur out of the blue; as we will see, it's important to monitor how pressure is building up by carefully measuring changes in the earth's crust. We'll then be able to trigger the alarm as soon as the ground starts to tremble.

Recent years have seen rapid progress in understanding early warning signals of critical transitions.[11] Quite diverse systems appear to have common features when approaching a tipping point. For example, systems may start oscillating between alternating states. This can be observed in lakes before they shift to a turbid state. Another type of behavior is a so-called increase in autocorrelation of signals. This means that systems recover slower from a perturbation as they approach a transition. This can, for instance, been seen from climate records that date back 34 million years. At that time, the tropic era of our Earth drew to an end. Just before this so-called greenhouse-icehouse transition, the atmospheric composition became more and more constant. A third type of behavior arises because systems are more susceptible to perturbations when they are close to a critical transition, which may give rise to an increase in variance. This may occur in brain signals before an epileptic seizure, when the electric signals from the brain show wild jumps.

In really complicated systems, it is not always clear what precisely we need to measure to pick up these early warnings. In every situation, you have to find the right indicator. And it is important to have detailed measurement with high spatial and temporal resolution. To probe critical situations, it's often useful to go to places where the pressure is highest. When studying famine, for instance, you go to Africa. If you're concerned about megacities, you go to India or Japan. Those are the places you'll see parameters beginning to shift. Many such parameters that we encounter in this book are poorly monitored. We need to measure on a much shorter timescale and with much greater accuracy than we do at present. Accurate measurement enables you to respond in time—a notion that is self-evident in the world of technology. Modern fighter-bombers, for example, are utterly reliant on it; without their sensors and control systems, they would become so unstable they'd simply drop out of the sky. Much the same is true in many other fields. If the outbreak of a new contagious disease in an isolated village is detected quickly enough, it's possible to prevent it from spreading. Rapid adjustment can restore stability.

In addition to more accurate measurement, we should try to use technology to delay transitions. One way of doing so is decoupling the global networks of interactions. These networks have become very tight, as we have seen, as a result of the human effort to increase efficiency. So one solution would be to undo this effort. We then need to decrease efficiency, diminish connectivity, and slow down speed. This is what Thomas Homer Dixon and others propose. This uncoupling is, as he remarks, difficult to attain because it also means giving up some of the gains that increased efficiency have brought. We would be paying the price of a regress to avoid global crises. This is certainly a correct observation in many areas. For example, it is difficult to see how financial stability can be attained without adding extra friction (see chapter 5.5). But there is an increasing number of fields where new technology makes decentralization possible without a loss of efficiency. Chemicals, for example, may be produced on a smaller scale than hitherto possible (chapter 2.5). In addition, communication networks and the distribution of electricity can be made efficient without increased vulnerability (chapters 3.2 and 2.2). What counts is the structure of the networks and the nature of the coupling mechanisms. More insight in complex systems may lead to new strategies to cope with the problem of globalized crises. That may give us the elbow room we need to keep on top of nonlinear dynamics.

Measures to uncouple global networks and postpone crises should preferably be based on existing proven technology. Bringing in untried technology risks creating new feedback mechanisms that we can't entirely oversee and that could make matters worse. What's more, it takes valuable time to establish a new technology, which further narrows our window of opportunity for avoiding catastrophe. Existing technologies offer a much faster and safer route. A good example is the control of electricity networks. In Europe and the United States, these have grown so intertwined that the failure of one power plant can have a serious knock-on effect. A single interruption can easily trigger the breakdown of the entire infrastructure. To stabilize the electricity grid, interdependencies of this kind will need to be reduced. If every power plant were to limit its services to its own region, only surplus power would have to be transported elsewhere. Loosely coupled power plants like this wouldn't threaten to bring each other down. Another possibility is to make the grid itself more flexible by giving it the ability to reroute power. Rerouting is also being pursued as a strategy for telecommunication networks. In both cases, the solution involves established knowledge of distributed control.

All the same, it's important to continue to develop our science and technology. Deploying existing technologies—although preferable in the first instance for the reasons just stated—may prove insufficient. We know, for example, that our fossil energy reserves will run out one day no matter how efficiently we use them. They are simply a finite, nonreplenishable resource. At some point, a transition will inevitably occur to other energy technologies. We will have to come up with something completely new; a paradigm shift will be needed so that we can establish a totally different energy infrastructure. It isn't easy to identify which new technologies have the potential to force a breakthrough in critical areas like this. Different possibilities often need to be explored in parallel. In most cases, it's perfectly possible to state the likely hurdles that need to be overcome. When considering hydrogen as an alternative fuel for transportation, for instance, we know we'll first have to solve the problem of how to store it. This book will seek to identify critical challenges of this kind.

We are also keen to stress throughout that this isn't all just a matter of technology. Humans play a key role. In many cases, technology is readily available but is difficult to implement for societal reasons. Our society has its own collective dynamics, which may be hard to change. We are currently in a stable state of overconsumption of fossil fuels, for instance: Cheap oil has made it prohibitive to work on alternative technologies, and so it will be difficult to move toward another stable state. For that reason, we will also focus on the societal acceptance of the technologies featured here.

Only a few experts that speak out in the remaining parts of this book are complexity scientists as such. Most scientists we will meet just hit complexity in their particular field. They use a common language with respect to transitions and stability. They also share a common concern of the problems of our time. The challenges that lie ahead are extremely serious. There are many themes that must be addressed as a matter of urgency by anyone desiring to make the world a more habitable place. It is clear, though, that the timely development of alternatives will make any transition less abrupt and could therefore help guide us through what promises to be difficult times.

Part 1
NEEDS

1.0

VITAL NETWORKS

The explosion in the world's population appears to be slowing down. Fifty years ago, an average woman had between five and six children. The global average now is just 2.6. In a mere two generations, therefore, the reproduction rate has sunk to slightly above the replacement level, which is currently 2.3. In half the world, people are having fewer children than needed to maintain the species. This includes countries like the United States, China, and Indonesia. In the European Union, Japan, and Russia, the population is shrinking for the first time in human history for reasons other than war, disease, or other calamities. It is a matter instead of free will, with women's education and rising prosperity helping produce a remarkable slowdown. Worldwide, the number of births is no longer increasing, which gives reason for optimism. The human race is incredibly flexible when it comes to procreation. Only in certain strongholds of Islam and Christianity—along with much of Africa—are birth rates still well above replacement level.[1]

The overall world population continues to grow, however, because lots of countries have relatively young demographic profiles. Life expectancy is rising in many places, too. Our species is currently growing by 75 million a year, which means we'll need more food, water, and housing in the future. The real problem, however, is that wealth has been growing at a much faster rate than the population. Many nations are undergoing rapid economic development, which is in turn changing patterns of consumption. People have begun to eat more meat, use more dairy produce, and consume more energy. We now live in a world where more children are obese than are underfed. With demand exceeding supply, it is invariably the poorest who suffer. China and a number of Arab countries are already buying up huge areas of farmland in Africa to secure their own food supplies.

How can we reduce the amount of waste in wealthy regions while simultaneously securing food and water for those who have no choice in the matter? Can we cope with increasing prosperity? The amount of fresh water available per head of the world population is already just 25 percent of what it was in 1960. One in three people face water shortages.[2] What's more, no water also means no food. Unless the most deprived populations cease to be subjected to this intensifying stress, we are likely to see increased migration and more frequent conflict. In all probability, the roots of the 1994 massacres in Rwanda were not racial but a matter of food shortages.[3] We have been warned.

Technology enables us to do more with our planet's finite supplies of water and farmland. This has been the case ever since Thomas Malthus's famous prophecy of demographic catastrophe. The potential for introducing new technologies has by no means been exhausted. In the past, however, the introduction of new technology often led to increased dependence. As farmers begin to use new cultivars, fertilizer, machinery, and irrigation, they become enmeshed in a network of suppliers that tightens and becomes more vulnerable to change as efficiency increases. A highly productive farmer is far more susceptible to an increase in the oil price or the appearance of a new pest. Thus, intensifying farming increases the risk of food crises—the real danger posed by our attempts to drive up food production. We saw this for the first time on a global scale in 2007.

In the following chapters, two experts show how we can increase food production and improve the provision of water while preserving flexibility and adaptivity. Frank Rijsberman, a water expert who now works at Google.org, explains how rural communities can use new communication techniques to make them less dependent on their water companies (chapter 1.1). Monty Jones, a plant breeder from Ghana and worldwide food ambassador, shows how the diversity of African agriculture can be a strength (chapter 1.2). The two experts also demonstrate how new ideas about flexibility and adaptivity will enable us to prepare more effectively for climate change and, with luck, deal more efficiently with political troubles. Their approach has the potential to meet basic food and water needs well into the future.

The examples in this part of the book focus on regions where shortages are felt most acutely. But improving the global food situation isn't only about farming in faraway places. We can also achieve a great deal by focusing on how we consume our food. Many people don't need all the calories they take onboard each day. Westerners certainly eat too much and throw away too much, which is why altering our eating patterns will be another important way to make better use of agricultural output. The U.S. Department of Agriculture and the Food and Agriculture Organization (FAO) estimate that Americans dump 30 to 40 percent of their food uneaten. Two-thirds of what Americans throw away consists of fruit, vegetables, milk, cereal products, and sugar. The corresponding figures are high for Europe, too.[4]

Food is wasted in shops, restaurants, and in people's homes. Eating out is a major contributor in terms of discarded calories. In the United States, one meal in two is now eaten outside the home, with one in three consumed in the car. Prepared food is often highly perishable, making it more likely to be thrown out. The relative amount of prepared food we consume is growing steadily and not just in America. An increasing proportion of the world's population is living in cities, which means more and more people will rely on prepared food. In cities less people eat at home.

The challenge will be to keep the number of calories at a healthy level while ensuring that we waste much less. Reducing the perishability of food might help in that respect. It could have a similar impact on the world's food supply as improving irrigation or increasing the flexibility of production. Crucially, the problems we face in terms of food and water supply are mostly related to a wrong attitude rather than overpopulation.

1.1

WATER FOR LIFE

Over a billion people don't have access to a safe water supply. And a third of the world's population lacks basic sanitation with the result that more than 2 billion human beings are afflicted with infections that result in diarrhea and other diseases. Tens of millions of them die every year.[1] Improving this state of affairs poses a massive challenge. Take sanitation: What if we could provide basic facilities for all those people over the next 20 years? You'd have to hook them up to the sewer system at the rate of half a million a day. We know how to install individual toilets and sewage pipes, but a project on that kind of scale is way beyond our capabilities. It would not only require new technology but a huge amount of money and political will, too. The challenges for providing all humanity with access to clean water are similarly gigantic.

It's not a matter of scarcity. There is enough drinking water for everyone on Earth even as its population continues to grow. According to the United Nations, a human being needs 20 liters of drinking water a day to live healthily. Every year, 100,000 cubic kilometers of rain fall on the earth, which translates into 40,000 liters per person per day. That would be plenty even if you only manage to tap a tiny fraction. Sufficient drinking water is available for all even in the driest regions of the earth. The problem is one of quality: People don't die of thirst; they die from drinking water that's not safe.

The use of water for agriculture is another story. Roughly 70 percent of the human use of fresh water is for farming. People rarely realize just how much water agriculture requires. It takes 1,000 liters to grow the wheat for a single kilogram of flour, for instance. Other products soak up even larger amounts of water. A kilogram of coffee needs 20,000 liters, and a liter of milk takes 3,000—mostly for the cattle feed and the grass consumed by the cow. And you need as much as 35,000 liters of water for every kilogram of grain-fed beef you produce. Our current Western diet means we consume around 6,000 liters of agricultural water a day—a substantial proportion of which goes into the cultivation of animal feed. The average Indian needs only 3,000 liters a day, primarily because of that country's large number of vegetarians.

In principle, the huge amounts of rain that fall on the earth mean that even those numbers should be sustainable. The problem is that global rainfall is so uneven. A quarter of all continental rain falls on Canada, while other regions can go for a year without a single drop. And even when there is enough rainfall every year, much of it falls in the space of a few weeks during the monsoon season and then flows away again immediately. For agriculture, therefore, the water problem is more a question of regional shortages. Hence, there are two different water issues. The problem facing agriculture is how to store and share a scarce resource; for drinking water, it is a question of affordable access and sanitation.

SHORTFALL

The shortfall in agricultural water is already making itself felt. According to the United Nations, only a few small countries—Bahrain and Malta among them—suffered from a serious shortage of water in 1950. Today, the list runs to twenty major countries, including Kenya and Algeria. Running into the limits of our water reserves will make it hard to cultivate more food. Water scarcity also entails severe environmental degradation. This can plainly be seen in India, where people have taken to pumping groundwater on a massive scale—a process encouraged by subsidized electricity and plentiful opportunities to siphon off power illegally. A million new pumps are added every year, and a similar number of farmers are sinking new wells and flooding their fields once or twice a week. Two-thirds of Indian farmland is now irrigated using groundwater. This has enabled the country's agricultural output to keep pace with its population growth, thereby avoiding a Malthusian catastrophe. However, even though water specialists tend to view groundwater as a renewable resource, it is currently extracted faster than it can be replenished. India is in a race to the bottom, pumping twice as much water as it receives in rainfall. In many places, the water table is falling by 10 meters a year.

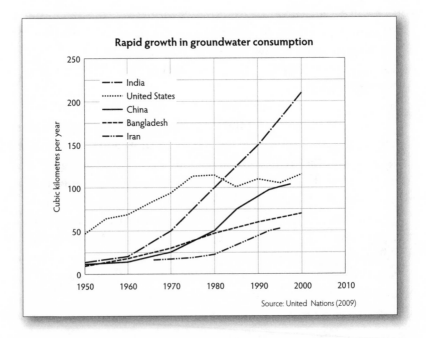

Rapid growth in groundwater consumption

Source: United Nations (2009)

The water table is falling by as much as 10 meters a year in some countries. Rather than installing ever more powerful pumps, we should try to reduce waste of water. The average global efficiency of irrigation is just 30 percent even though more than 90 percent efficiency has already been achieved using better technology. *Source*: Water in a changing world. *United Nations World Water Development Report 3*. United Nations World Water Assessment Programme (WWAP), 2009.

People often have to drill hundreds of meters before they reach water, and well after well is running dry. Within a generation, it will all be gone. Groundwater is being depleted in other parts of the world, too: Problems are already occurring in Peru, Mexico, and even the United States.

Rivers are drying up as well. Lots of major rivers around the world no longer make it to the ocean. The Ebro in Spain, for instance, is famous for the rice paddies that thrive in its delta. However, farmers upstream now extract so much water for irrigation that the river is drying up, and the delta is salinizing. The Murray River in Australia has dried up because of overexploitation by cotton and rice farmers, threatening Adelaide's water supply. Some of the world's biggest rivers, including the Yellow River and the Colorado, are no longer reaching the sea. Other rivers are also losing water.[2]

IMPROVING IRRIGATION

The shortage of water will be exacerbated as irrigation is applied to the cultivation of an ever-greater volume of crops. It will remain a key issue, therefore, in any technique designed to improve agricultural yields. A sixth of the world's farmland is irrigated, and that land provides 40 percent of all our food.[3] Consequently, any further intensification of farming inevitably means increased water consumption. "We have the technology to make the irrigation process a lot more efficient," says Frank Rijsberman, program director at Google.org—Google's Mountain View, California–based philanthropic arm. As a former director general of the International Water Management Institute—a leading research organization in Sri Lanka—Rijsberman's experience with water issues goes back a long way. We talk via a transatlantic video link that cancels out the tens of thousands of kilometers separating us. It feels as though we're sitting on the same red sofa beneath rows of Post-it notes stuck on the wall.

"Water management experts tend to think in terms of water supply. When demand increases, they try to engineer new water resources. But you can't create new water. It circulates in a cycle. You can only divert a certain proportion of that cycle for human needs. So you have to focus on demand as well." Increasing the efficiency of water use can reduce demand. The average global efficiency of irrigation, for example, is about 30 percent. Israel, by contrast, can achieve 90 percent efficiency. There are much more effective methods than simply flooding fields, which allows most of the water to evaporate or drain out of the soil. Sprinklers are twice as efficient, and drip irrigation even more so. There is also room for improving the irrigation channels themselves: Irrigation systems often feature open channels, from which you get a lot of evaporation, causing the remaining water to silt up.

The challenge is to make irrigation technologies cheap enough for the poorer parts of the world. A simplified drip irrigation technique has been successfully developed in India, initially for the cotton fields of Madhya Pradesh and Maharashtra. These use so-called Pepsee kits that consist of perforated hoses placed in the soil to deliver water and, where appropriate, fertilizer. It's a simple setup, but it doubles yields and halves water use. "There's still plenty of leeway for reducing water consumption if you use technologies of this kind. It's like energy conservation—you have to persuade people to do it. Information exchange is crucial," says Frank Rijsberman. More effective water management does indeed rely on the support of a local community. In India, rural communities traditionally formed around the management of irrigation systems. Nowadays, however, many of those communities have ceased to function properly, and their breakdown has adversely affected irrigation, too. Local initiatives have been launched in India to collect and store monsoon rain. Villagers are digging channels, reservoirs, and cisterns to prevent rainwater from draining away unused. Restoring old water collection systems will result in a more continuous water supply.

"But the potential for improvement is greater still," Rijsberman thinks. "Irrigated farmland is usually watered 100 percent of the time. As soon as we've built an irrigation system, we start acting as if rain has ceased to exist. That's certainly the easiest solution: You simply have to open the valves once. It's what I was taught at university, too, when I was studying irrigation technology. Rain is unreliable as a water source, so it was deemed unwise to depend on it. But 60 percent of all rain is absorbed directly into the soil, where it can be used by plants. It's silly to ignore that. There are many places where you only need a little supplementary irrigation, provided you do it at the right moment. But we don't yet know how to control a variable system like that with sufficient accuracy. Very little research has been done into this concept. For detailed control, you need fine-grained data about precipitation and fine-grained weather forecasting. At the moment, the weather data for many regions are notoriously poor. In Ethiopia, for example, there are 1,600 weather stations, only 16 of which release their measurements via the Internet. Most of them report on paper, which usually takes 2 months. By that time, the plants in the field could be dead."

To counter these problems, Google.org supports a project which—together with UN Habitat and a number of other partners—seeks to collect more detailed information about weather, water, and sanitation. "Think of mobilizing people in remote villages to text or email their data," Rijsberman explains. "That gives a parallel circuit of information exchange outside the official government figures. The current goal is to get fine-grained statistics. But the concept can be expanded. Farmers who send weather observations, for instance, could receive advice in return. The resulting weather data could then be used for the detailed control of irrigation. You can also ask people to report cases of disease. We now have a cell-phone app called 'An outbreak near you,' which gives you the means to become part of a giant disease surveillance system and eventually nip an epidemic in the bud. It's a completely new way of dealing with problems like that." Not to mention a further example of the growing role and importance of communication networks. The impact of communication networks is in this case, like in many other examples, that users also become a network and share knowledge and ambition.

Another promising idea is to link irrigation with sewage. "Wastewater contains valuable manure and other growth substances," Rijsberman confirms. "If farmers have to choose between freshwater and sewage water for irrigation, they often prefer the latter. And sewage water is always available, whereas irrigation systems are frequently interrupted. Farmers are even willing to pay for sewage water. That would solve a major problem because in many cases, nobody else can afford to pay for sanitation. So it can be a good idea to let farmers take care of it. They obviously have to take precautions when using sewage water. But in most cases, wearing boots is all it takes to protect them from infection. And their produce has to be safe to consume as well, of course. The problem is that all this has to be organized between different parties who wouldn't normally be in contact with each other."

COMMUNICATING VESSELS

It is clear from the latter example that the issues associated with the provision of drinking water, sanitation, and agricultural water are closely linked. These streams are communicating vessels—literally so, in many cases. Solving a sanitation problem can often help achieve a safe water supply, too, just as diverting too much water to agriculture can endanger the supply of drinking water to a city. That's why it makes sense to organize improvements in the water supply on a small scale, within the capillaries of the network, as it were. At neighborhood or farm scale, it is still obvious that the different water streams are interconnected.

"Water purification has traditionally been viewed as a task of the government and one that could only be carried out at large plants. But that's now changing," Frank Rijsberman says. "There have been exciting breakthroughs in purification technology, which means we now have membranes that can be operated on a small scale. That's given rise to a new class of small water treatment devices. These filter water and often use UV light to kill germs. They can be really small, and you can order them through the Internet for a couple of thousand dollars. Small-scale purification units like this are starting to appear in places that hitherto didn't have clean water. A new industry is already growing up around them. Thousands of small shops have begun to produce and sell filtered water in poor parts of cities in the Philippines and Indonesia that have never been connected to water networks. It's a lot better than sending out huge water tankers to places where an acute need has built up. By the time you've distributed that water, it's often no longer reliable for drinking. These small-scale devices solve that problem. Membrane technology is progressing rapidly, so the price should come down further. I also expect nanotechnology to lead to even cheaper and more refined purification devices. Small floating nanoparticles, for instance, have the capability to remove certain contaminants selectively. Nanotechnology could also tweak the filters so that they consume an order of magnitude less energy." This recalls the pattern of development we are witnessing elsewhere in industry. Chemical plants, for instance, are evolving toward small distributed production units (see chapter 2.5).

"The water purification devices are a further development of the technology for desalination that is now used routinely to produce drinking water in hotels and in cities like Singapore. They use similar membranes for reverse osmosis. Desalination is still too expensive for agricultural use, but when prices come down, the technique might become attractive to farmers, too, provided they combine it with the most efficient irrigation techniques. That would be very exciting, as it would solve a key problem. Such an advance would liberate farming from its dependence on rain, river flows, and wells." Tapping water from outside the fresh water cycle would create a much more reliable source.

New crops could further reduce agriculture's thirst for water. We will see in the next chapter how new varieties of rice can be bred that require less water and are more resistant to drought. Frank Rijsberman is convinced that we'll see more developments of that kind. "New cultivars and new farming methods could drive down the water requirement per kilo of rice from 2,000 to 500 liters. Traditionally, plant breeders could not optimize for water consumption because it was too difficult. They bred varieties that optimize for a single trait, such as sensitivity to a specific pest. But drought resistance depends on lots of different traits which you have to optimize simultaneously. Biotechnology has now provided the means to identify the desired traits more accurately. We now know the entire rice genome, which means that more complex multitrait optimization is now feasible."

A wide array of techniques is therefore available that could reduce the stress on water systems. They are urgently needed as climate change is likely to add to that stress. "Adapting to climate change is all about water," Rijsberman says, "not because it's going to get warmer but because climate change will bring more extreme events, such as droughts and floods. There will be much more variability. The good news is that we don't need to develop new approaches; we can tackle those issues using the techniques I have just outlined. So it will be even more important to achieve a fine-grained understanding of the weather system," he continues. "We will have to make better use of the cloud of people armed with cell phones and Internet to keep track of the changes. That will help us introduce improved agricultural techniques, such as irrigation, and make better decisions with regard to crops."

One way of overcoming irregularities in the water supply is to improve storage, Rijsberman points out. "Here in the United States, we have dams with reservoirs that together store 5,000 cubic meters of water per inhabitant. In Ethiopia, it's less than 50 cubic meters. Even today, that's not enough to guarantee a continuous water supply. We're unlikely to solve the water supply problems in Africa without new dams." The pros and cons of dam building are examined in chapter 2.3 in the context of energy supply, which shows that dams don't always yield a net environmental benefit. All the same, Rijsberman thinks there are still environmentally sound opportunities for building new dams, especially in Africa.

It's the only point in our discussion where he proposes a large-scale solution. As a child of the Appropriate Technology movement, he's a firm believer in small–scale bottom-up approaches that are close to people's actual needs and don't require government intervention. "But small-scale dam projects often run aground due to mismanagement and poor use of technology. Then shallow reservoirs can become sources of infectious diseases. Managing a dam isn't easy, and there are advantages of scale. At the same time, large dams create a risk of corruption and abuse. To deal with that, good governance requires that you'll have to open up the budgets, be transparent about contracting and pricing processes, and ensure that the water users themselves are closely involved in projects of this kind."[4]

In many other areas of water management, decentralized technology will improve the delivery of clean water and sanitation to poor communities. Although governments invest in large-scale programs such as dams, there is much to be gained through smarter consumption and small-scale technology. We need the low-cost technologies that Rijsberman mentioned, such as off-the-shelf treatment plants, available over the Internet, capable of reaching remote areas and slums. In time, this will stop our rivers from drying up, salinizing, and becoming polluted. It could also allow water tables to rise again and diminish the number of disputes over water allocation. And most of all, it will hold out the prospect of a healthier life to hundreds of millions of poor people around the world, who currently lack access to clean water.

1.2

FOOD FOR ALL

Can we feed the world? Although the rate of increase is falling, the world's population continues to grow at an explosive rate, doubling since the early 1960s. Fortunately, the quantity of food has increased even faster. The average human being in 2010 has 25 percent more food than in 1960 despite the huge increase in population. Although that's an average figure, the proportion of humanity that is undernourished has also fallen. Extending and intensifying has improved our fate. But we have reached a point where not much more nature can be converted into farmland without serious negative impacts on other vital environmental services such as water catchments, carbon sequestrations, and conservation of biodiversity. A quarter of the world's ice-free surface is already used for farming. So how can we increase food production to feed the increasing number of people as well as improve the nutrition of the millions who are still malnourished, especially in Africa?[1]

Pessimists regularly predict catastrophic food shortages. The specter of starvation can be traced back to the ideas of the English parson Thomas Malthus, who fretted about the population explosion he witnessed toward the end of the eighteenth century. A typical couple at the time had four children and sixteen grandchildren, which meant the population was growing exponentially. Malthus anxiously predicted a shortage of food, as he didn't believe new farmland could be cleared fast enough to keep feeding all those extra mouths. Linear growth in the area under cultivation—and hence, the production of food—was the best that could be hoped for. Something would have to give, and Malthus was convinced that humanity would be stricken by genocidal war, plague, and other epidemics. Starvation on a massive scale would then restore the balance between population levels and food supply. Yet the catastrophe he predicted never came about.

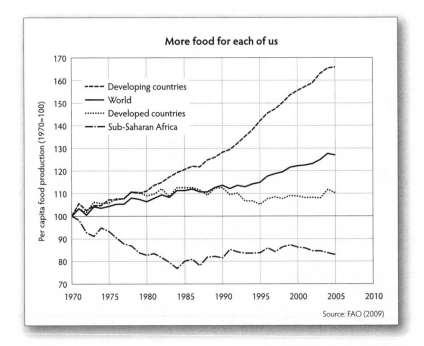

The world's population continues to grow rapidly. Fortunately, the food supply has expanded even faster in most parts of the world. Overall, there is more than enough for all of us. The fact that shortages exist is a question of politics, communication, and knowledge transfer. All these factors are now so intertwined that any disruption to the food supply tends not to be a local matter anymore but to strike globally. *Source*: The state of food and agriculture, FAO. The diagram is a combination of data from several of these yearly reports.

This progress is only partly attributable to opening new farmland. It has grown so little that this factor alone would never have been enough to go on feeding our growing population. If increasing the area under cultivation was the only way to raise food production, Malthus would have been proved right. Fortunately, improved agricultural techniques have substantially increased output from the existing land.

ADDING NITROGEN

The development of synthetic fertilizers was an important step. They add nitrogen, phosphorus, and potassium to farmland—a process that now occurs on a formidable scale worldwide. To produce fertilizers, humanity now extracts more nitrogen from the air than nature does. In doing so on such a huge scale, we have effectively seized control of the planet's nitrogen cycle, which means that for the first time in Earth's history, we have become the dominant factor in the natural cycle of a basic element. Humans have been fixing nitrogen faster than nature since 1980.

The process we use for nitrogen production is basically unchanged since its discovery in the early twentieth century. We still manufacture fertilizers at pressures between 150 and 250 atmospheres, a temperature between 300 and 550 degrees centigrade, and using a considerable amount of energy. Around 2 percent of the world's total energy consumption currently goes into the production of fertilizers. The nitrogenase enzyme, by contrast, can achieve the same result at ambient temperatures. There are several bacteria that use this enzyme, fixing nitrogen in the soil in the process. Most of the nitrogen that is naturally present in soil comes from the *Rhizobium* and *Bradyrhizobium* bacteria. It would be a huge advance if we could imitate them and learn how to produce our fertilizer under more favorable conditions.

Other technologies were equally important in avoiding a Malthusian catastrophe. Irrigation has doubled or tripled agricultural production in many regions, and machinery has doubled it again in many more. Pesticides and improved cultivars have further boosted production, and genetically modified (GM) varieties may continue the pattern. In many regions but not in all, these techniques have hit limits. Irrigation feels the scarcity of water and increased competition from industry and households. Pesticides have their limits, too.

It is equally important to note the changing pace of these successive revolutions in agriculture. Irrigation dates back almost to when human beings first chose to settle and cultivate their crops. As late as the 1950s, however, a mere 3 percent of all cultivated land was irrigated compared to a figure nowadays of 20 percent. Machines like tractors and combined harvesters appeared in our fields slowly, allowing productivity to be scaled up without a corresponding increase in labor, although even today there is farmland in the European Union that continues to be cultivated without the use of machinery. Synthetic fertilizers, meanwhile, have been possible since the first nitrogen plant was built in Germany in 1913. Their use didn't take off, however, until the 1950s. Nowadays, though, the pace of innovation is increasing. New cultivars were systematically introduced in the second half of the last century, and the application of biosciences, including genetic modification, has only been possible since the 1980s.

THE INNOVATION CYCLE

Each new revolution is more rapid than the last, not least because it can build on the previous one. The increase in scale permitted by agricultural machinery created a need for new pesticides, and improved cultivars often demand so much water and nutrients that they can't grow without fertilizers and irrigation. The complexity and interdependencies of the enabling organizations also rise with each successive revolution. An example is the development of irrigation. In the ancient Nile region and the Euphrates and Tigris basin, the high level of organization demanded for the control of water contributed to the rise of the first regionally powerful states. Likewise, the introduction of synthetic fertilizers takes coordination on a global scale. Africa, for instance, produces less fertilizer than it imports. The international network is even tighter when it comes to the use of new cultivars. Seeds of hybrids that cannot be produced on smallholdings must be bought and often involve complex production contracts. And with GM crops, the network is tighter still.

Thus, each new technology increases the level of global interdependence, and the advances in agricultural science have made the worldwide food network progressively denser. The resultant couplings render the global food supply increasingly sensitive to fluctuations in prices and supplies of key inputs. When energy prices soared in 2007, the cost of fertilizer and transport rocketed with them, adding to the cost of food production. Similarly, because of the interlinkages, farmers cannot ratchet up food production the moment an increase in demand is detected. It not only takes an entire growing season, but it also requires changes in the complex input supply systems. That means there is little flexibility within the network to absorb fluctuations, especially when there are limited food reserves. Hence, there was a sharp increase in worldwide food prices in 2007. The more integrated you are in the network, the more susceptible you are for such effects. African farmers cultivating rice and vegetables for their own families were probably shielded to some extent from the turmoil engulfing worldwide food prices. But larger farmers weren't. When it comes to increasing food production, it is important to bear these interdependencies in mind. Should we continue to integrate global food production ever more tightly and with ever more complex technology?

Each new technology has also tended to favor increases in the scale of production. That's why agriculture steadily grew in scale throughout the last century. Large processing plants have developed because they should be cheaper and more efficient. A similar argument applied to farmers. But increased scale in farming makes it progressively harder to adjust and to provide the detailed care and attention that a smallholder provides. The African tradition to combine different crops on one field offers food security in unpredictable seasons. This is not possible in large-scale systems that favor monocropping. If everyone around you is growing coffee, you can't suddenly switch to sunflowers to escape poverty. A hundred kilograms of sunflowers aren't going to keep a local factory running. Once a farmer gets locked into the system, it's extremely difficult to do anything else. It can sometimes take a whole generation to switch to a different crop.

Moreover the belief in increasing scale is not always borne out by the statistics. Data produced by food manufacturer Unilever suggest that small plants can be every bit as efficient as large ones. The company could find no link between scale and costs. In many cases, it is theoretically possible to produce more efficiently if you increase your scale, but big installations have great difficulty when it comes to optimizing their processes. A small plant can be adapted more quickly and can also eliminate by-products more easily.

THE ABUNDANCE OF AFRICA

Those were our considerations when we discussed the topic with Monty Jones, an agricultural scientist from Sierra Leone, who is now based in Ghana as executive director of the Forum for Agricultural Research in Africa (FARA). *Time* listed him in 2007 as one of the world's hundred most influential people for his role in seeking to free Africa from the grip of famine.

"Africa has the potential to feed the rest of the world," he says. "In fact, Africa *must* feed the world." We met over lunch in the garden of a hotel in London to which Jones has traveled to discuss the global food situation with British parliamentarians. "This is basmati rice," he says as he samples his curry. "Probably from India. It's fluffy. In Africa also, we prefer sticky rice." As a plant breeder, Jones knows all about rice, having crossed the two principal varieties used as cultivars around the world. The Asian *Oryza sativa* is high-yielding thanks to its many secondary branches, all of which can bear grains. The African variety *Oryza glaberrima* is not branched and so has a significantly lower yield. On the other hand, it has been cultivated for more than 3,500 years and has developed protective mechanisms that enable it to adapt to drought, diseases, and insect pests—all of which are major constraints in Africa.

Combining the two strains required a painstaking process of crossing, testing, and backcrossing the offspring with one of the original species. "Most interspecific hybridization results in sterility," Jones explains. "Many people had tried to do the same thing." It was the patience of Jones's team, combined with the ability to test the results under local conditions, that brought success. It couldn't have been achieved anywhere else in the world, as no one was commercially interested in varieties that grow better under African smallholder conditions.

The resulting "New Rice for Africa" ("Nerica") had a surprise up its sleeve: In African rain-fed conditions, it outperformed its African and Asian parents through a phenomenon that plant breeders call *heterosis*. Nerica has a lot of secondary and tertiary branches and 50 percent more grains than its high-yielding Asian parent. It also grows faster, allowing for two or three crops a year. "Even in traditional agriculture, without fertilizers or irrigation, the yield is double that of the African variety," Jones says. But Nerica retains important traits that enable it to withstand the rigors of the African continent. It can, for instance, survive without water for 2 weeks. It also has a higher protein content: "I'm talking about the survival of children. In some regions, people eat rice three times a day. With a higher protein content, they can get a more balanced diet simply from rice, maybe with a little fish."

This is a promising avenue for further crop improvements. "We should go on researching. There will never be a point at which you can say 'this is enough.' The African population is the fastest growing worldwide and urbanization is also more rapid than anywhere else. This means that more people have to be fed from less land. Africa must therefore do everything it can to increase yields per unit area and per unit input. But more important still is the need to develop crops that can withstand major emerging stresses as climate change intensifies, aggravating droughts and floods, and encouraging diseases and pests." Jones also thinks it's vital to further raise the level of protein, beta-carotene, and other micronutrients in the cultivars. "It's not only yield that's important; science ought to focus strongly on improving the nutritional value of crops. The poor should be able to feed on rice as the staple food they can afford and still grow strong and healthy and reach their full cognitive potential, which they require to pull themselves out of poverty."

ROADS

Although the popularity of Nerica rice is growing sharply, Monty Jones believes that more breakthroughs will be needed in rice and other crops before Africa can feed itself. "Southern Sudan is very fertile. It is a high potential area for food production, and it is estimated to be capable of growing enough food for the whole of Africa. There are similar spots elsewhere. The continent has 14 million hectares of inland valleys with plentiful water where you can reliably grow crops. But only a million hectares of these are presently used as farmland. We have the technology, but a lot of other resources are needed to develop these regions. For example, farmers in southern Sudan who may be willing to produce a surplus could not sell their extra output. The roads are so bad that fertilizers can't get in and the harvest can't get out. It's cheaper to ship grain from Australia to Mombasa (Kenya) than it is from southern Sudan, which is just next door. We need roads, markets, and reduced tariffs. If we succeed in these things, we could indeed be the breadbasket of the world."

If there is such large potential for agriculture, would it then be a good idea to grow crops for biofuels? "I strongly feel that biofuel production often competes with food production," Jones contends, "especially when there are subsidies for biofuels which are incentives to divert resources from food production. Ethanol should never be made; it costs you 1.5 times the energy that it provides and in its production does not have much better greenhouse gas emissions." There is also a close relation between energy costs and food production, he remarks. "Oil prices affect agrarian production in many ways. When fertilizer, machines, and transportation all become more expensive, seed prices also go up. When food prices rose sharply in 2007, it put an extra burden on many farmers, and some didn't have enough cash to buy new seeds. That reduced the area cultivated. So when there is a crisis, production goes down, and the crisis is further aggravated. That affects all."

Would the production go up if genetically modified cultivars would be better available for Africa? Monty Jones is cautious. He is well aware of his political role in the food industry. "In addition to the important conventional applications of biotechnology, GM crops probably have a useful role because there are certain traits that we cannot achieve through conventional breeding. They may be important, for instance, in raising food quality to improve the health of people. The safety of humans and the environment is of paramount importance, but with that caveat, we should keep open as many options as possible to be able to respond to future food demands, diseases, pests, and environmental challenges that we cannot predict."

But the cultivars should be developed in Africa itself, he thinks. "The use of advanced breeding techniques should not be at the expense of Africa's increased dependence on non-African breeders and multipliers. You should be well aware that the green revolution, with its improved crop varieties, irrigation projects, and synthetic nitrogen fertilizers, has yet to reach Africa. There were never enough resources devoted to it, partly because of the huge diversity in Africa. We grow rice, maize, plantains, bananas, millet, sorghum, and many other food crops, some of which (such as African leafy vegetables) are not grown elsewhere. And our production systems are also very diverse in political, social, and environmental terms. That's different from Asia, where 90 percent of farmland is devoted to rice, which is also the dominant staple food. It is easier to focus resources and implement improved agricultural practices when there is one dominant crop and production system. It is therefore a greater challenge for Africa to target technologies taking into account the continent's extreme diversity."

SELF-SUFFICIENCY

Education, Jones is convinced, is crucial for advancing agricultural development. "Africa is losing 25,000 professionals a year to greener pastures elsewhere. The continent is basically donating its intellectual capacity to the North. If it costs say $100,000 per person to educate to postgraduate level, it means Africa is contributing $2.5 billion toward the Northern Hemisphere's development without counting the opportunity costs of losing the benefit of their career contributions. This brain drain affects whole educational, research, and development systems. You can't have good education without well-qualified teachers and trainers. We have to keep our best people if we're going to have the innovation capacity needed to feed and employ our continent. We must, for example, continue to improve crops, to be able to cope with emerging challenges and opportunities, and there will never be an end to that."

This great diversity of Africa of which Monty Jones speaks is one of Africa's greatest advantages as we are entering an age when more flexibility is required and where we may decrease scale without loss of efficiency. Improved irrigation, fertilizers, and cultivars are certainly crucial to securing Africa's—and the world's—food supply. Worldwide, the potential of the green revolution is by no means exhausted. And while not repeating the harmful environmental effects, we should bring the benefits of the green revolution to African rural communities without closing off local options and tightening too much the international network of technological dependencies. For example, it is both vital and urgent for every agricultural region to produce its own nitrogen fertilizer. As long as Africa has to import most of its fertilizer, the continent will continue to be buffeted helplessly by storms in the international economy. Small production units, close to the end users, offer a more stable foundation for agriculture. We need to reintroduce flexibility in the food supply if we are to avoid dangerous shocks.

Part 2
EARTH

2.0

OUR PLANET

Mahatma Gandhi supposedly once said: "It took Britain half the resources of the planet to achieve its prosperity. How many planets will a country like India require?" Translated to the world order of today, his question would be: "What if China would aspire to the standards of living of the United States?"

Our planet is certainly flexible. A quarter of its surface has been plowed up, and its atmosphere, soil, and water have been fundamentally altered in many places. Humanity now extracts more nitrogen from the air than nature does, and we use more water than all the rivers put together. It's a miracle that Earth's systems have been able to withstand these interventions as effectively as they have. Many parts of the world are cleaner than they were a century ago. Pollutants like sulphur, nitrogen, and small particles are now routinely filtered from exhaust pipes and chimneys. We've mastered the problems of acidification and smog. But those were the easy tasks. The fact that we dealt with bad things in the past is no guarantee of a rosy future. Interference in our environment is too great for that. Humanity continues its assault on the planet. The toughest problems remain unsolved.

The truth is that we are already consuming more than one Earth can support. Just as a company can spend more than it earns by selling its assets, we are eating into Earth's capital, which was accumulated during thousands of years. In a report published by a group of leading scientists, it was concluded that we already have transgressed safe planetary boundaries in many respects.[1] We already have surpassed the carrying capacity of Earth's climate with a factor of 1.5, we are at a tenfold rate of bearable biodiversity loss, we extract four times more nitrogen from natural cycles than can be considered sustainable, and we are at the tolerable thresholds of the phosphorus cycle, ocean acidification, and stratospheric ozone depletion. Human civilization is out of kilter with the natural environment. We are using considerably more than one Earth.

Many subsystems of Earth react in a nonlinear, often abrupt, way. The world is heading toward a number of critical transitions. Our changing climate and the depletion of oil and raw materials mean that major changes should be expected. We are living in an overshoot, which makes a soft landing difficult to attain. The roots of the 2008 financial crisis were laid by people who spent $1.2 for every $1 they earned. How do we prevent a crisis when we already use up 1.2 Earth for the one planet we have? The pressing need for resources and fertile land has brought out humanity's darkest side. Wars have been started over oil. Hostile migration, sparked by diminishing grasslands, is one of the main reasons for conflict in Africa right now. If we can't alter our colonization of nature, we might find ourselves fighting each other on a global scale again. If we carry on as we are, exhaustion and extinction are possible. The major task facing us is to keep Earth's system stable.

The most pressing task may lie in the issue of climate change (chapter 2.1). Long before fossil fuels run out, we'll have to face up to the consequences of using these fuels. Global warming will be a much greater threat in 20 years than it is today. Changes in the atmosphere have never occurred as rapidly as they are now. Our current tools and social structures are not sufficiently effective for us to manage the climate or to prosper in hostile surroundings. We must either learn how to change the climate in our favor or develop technologies that will enable us to survive in different environments. Both are clearly lacking today. The development of science and technology in these areas should therefore be given the highest priority. If we manage to solve these problems in the decades ahead, we have grounds for hoping that our descendants will also survive into the distant future.

Energy is a key issue not only because of our changing climate. Our society needs energy to survive, to develop, and to prosper. We need it to prepare food and to provide comfort; to advance our science and technology; and to fuel our vehicles, telecommunication networks, and domestic appliances. The facts speak for themselves. No additional oil is being produced inside the planet. We may not know whether our reserves will dry up in 2040 or 2060, but we do know that they are finite. We look in chapter 2.2 at how we might postpone critical transitions and in chapter 2.3 at preparation for the period beyond. Technology issues are not limited to the problem of finding new sources of energy. The distribution and storage of energy are also key issues. Fuels must be transported over large distances, and electricity typically has an international network infrastructure. In both, we have so far benefited from economies of scale. Up to now, it's always been a question of bigger means more efficient and cheaper. But this has made the energy infrastructures more vulnerable and less easy to change. We'll analyze how to change that.

Consumption of raw materials is growing at a faster rate than the world's population. In two subsequent chapters, we explore how we can drastically reduce this in factories (chapter 2.4) and chemical plants (chapter 2.5). These chapters show the inertia of our industrial systems and how engineers can change that. They can, for example, learn a great deal from nature. Natural materials and processes are often better than those we could conceive ourselves. And nature frequently produces substances in a more sustainable way. New process technology will make chemical plants more sustainable, smaller, and more flexible. We will no longer have to hide plants away on isolated industrial estates; we'll be able to produce closer to the user.

Life has always left its marks on our one Earth; that's plain from the composition of the atmosphere. The oxygen we need to survive is highly reactive and readily bonds with carbon. Over time, all the oxygen would disappear from the atmosphere if Earth's vegetation wasn't constantly adding fresh doses. The atmosphere is an unstable mixture that is maintained by life on Earth. The planet and the life it sustains have evolved in parallel. Immense natural catastrophes have occurred, yet life and the production of oxygen have always prevailed in the end. There are serious grounds for concern. It is more than clear that human beings are having a huge and intensifying impact on our planet. Whether Earth would continue to sustain humanity is very much open to question. The human race didn't exist 200 million years ago, when—for unknown reasons—a huge quantity of carbon dioxide was released into the atmosphere and temperatures rose dramatically. Yet even a less severe disaster would put an end to human civilization as we know it today.

2.1

DEALING WITH OUR CLIMATE

We're standing by the observatory at the top of the Telegrafenberg (Telegraph Hill) in the German city of Potsdam. The neoclassical building towers over its surroundings. The hill is situated in the former German Democratic Republic, close to the place where the Berlin Wall once stood. Through the slight haze, we can make out the contours of Berlin and the smoking chimneys of power stations. To our right is another hill, the Teufelsberg, with an American listening post as a relic of the cold war.

Successive kaisers developed the Telegraph Hill in the nineteenth century, building a community of leading scientists there. Karl Schwarzschild used the telescope to produce his star catalog, the first in the world, while in the basement of the same building some 30 years earlier, Albert Michelson had studied light, measuring its speed and identifying certain inexplicable characteristics in the process. Albert Einstein worked here, too, basing his special theory of relativity on Michelson's discoveries.

"Fundamental natural phenomena have been isolated at this place," says Hans Joachim Schellnhuber, director of the Potsdam Institute for Climate Impact Research, which now occupies the brow of the Telegraph Hill. "For many years, scientists have withdrawn to the quiet of this hill to develop their ideas. My task today is to reverse that movement: Rather than isolating it, we want to bring knowledge together. And instead of withdrawing from the world, we have to engage with it—to make clear to people just where our climate is headed."

Schellnhuber has thrown himself into that task with considerable verve. He has been discussing scientific issues with German chancellor Angela Merkel, for instance. He knows that his climate message is a complex one, which is why Schellnhuber avoids statistically detailed predictions and focuses instead on a number of crucial "tipping points."

POINTS OF NO RETURN

"How much change can the earth sustain? Can we afford to allow the West African monsoon to collapse? Or the Himalayan glaciers to melt away? Will we be able to preserve the ice in the Antarctic? What happens if the Amazon rainforest disappears?" Hans Joachim Schellnhuber has identified a series of threats that, should they come to pass, would transform our biosphere irreversibly. Our climate system includes several positive feedback mechanisms with the potential to accelerate global warming, he explains. The melting of Greenland's snow and ice is one example. The white surface of that immense country reflects a considerable amount of solar radiation back into space. As the ice melts, however, Greenland is growing darker and is absorbing the heat of the sun more and more readily. This extra heating makes it unlikely that ice will ever reappear on this—the Northern Hemisphere's largest—island, making the melting of the Greenland ice sheet a tipping point—a point of no return.[1]

Another example is the Amazon rainforest, which currently retains a lot of water, creating a humid atmosphere that fosters a wealth of nature. If it were to disappear, the result might be a steppe to which it would be very difficult to bring back jungle trees and abundant water. Scientists refer to this complexity as a *bistable system*. The Amazon ecosystem is locked into one of two possible states: It is either rainforest or steppe, with no intermediate configurations between the two. "The Amazon rainforest may collapse by the end of the century," Schellnhuber warns. "It's not only the result of climatic shifts; illegal logging is also accelerating its disappearance."

Other worryingly close tipping points he has identified include the melting of the Arctic Sea ice cap. In the summer of 2008, the sea route north of Russia became ice-free for the first time in history. As the icebergs melt, the dark heat-absorbing surface of the sea is uncovered, further accelerating the warming process. Some scientists believe that the tipping point for the Arctic Sea ice has already been transgressed.

The disappearance of the Indian monsoon is a frightening prospect as well. Seasonal rain in India appears to be another bistable system, and continuing environmental pollution could lock the local climate into a dry state, inevitably causing food shortages in a subcontinent inhabited by more than a billion people. Or—no less terrifying—the

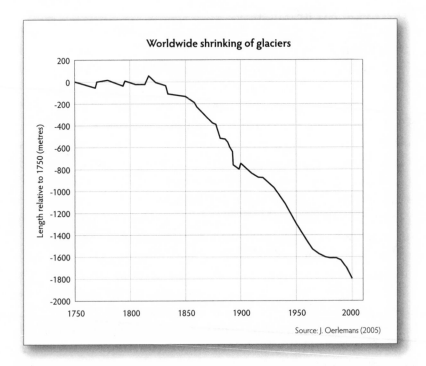

White glaciers reflect the sun's light, helping repel some of its warmth. As glaciers recede, the darker-colored earth beneath them will absorb more heat. This may contribute to the pace of climate change, making it harder for the glacier to reestablish itself. The challenge we face is to identify these tipping points in our climate and to learn how to avoid them. *Source*: Oerlemans, J. (2005). Extracting a climate signal from 169 glacier records. *Science,* 308(5722), 675–677.

monsoon could end up oscillating between these two stable states, with alternating dry and wet years that remove any possibility of adapting.

"Several of these tipping points appear to be close," Schellnhuber believes. "But their peak impact may be many years away. We need to be aware that if a tipping process is triggered, there is no stopping it. A good example is the methane trapped deep in the oceans. The amounts are huge. If it warms up, it will take a thousand years to bubble through sediments and water columns and reach the surface. But we won't be able to halt the process once it gets going. We're very close to the Greenland tipping point; if we go on emitting according to the 'business as usual' scenario, the system will tip by 2050 or so.

After that, it can't be stopped. The melting process would then take place within some 300 years, causing sea levels to rise by an additional 6 meters. Climate tipping points of this nature imply massive change."[2]

CRADLE OF CIVILIZATION

This is a completely new challenge in the history of humanity. Carbon dioxide (CO_2) levels had barely fluctuated since the rise of the first advanced civilizations—a stability which in a sense may be viewed as the midwife of our civilization. Greenhouse gases in the atmosphere function as a blanket, retaining Earth's warmth. Without it, planetary temperatures would be far below zero, denying human beings any chance to develop. Almost stable CO_2 levels allowed our ancestors to settle down and adapt to their environment.

Egyptian pharaohs and Babylonian kings obviously didn't go in for precise carbon measurements, but we can track historical atmospheric changes through the traces they have left in layers reaching down deep below the ice of Antarctica. Thanks to that record, we have a fairly precise idea of the pattern over the past 650,000 years.[3] The climate has remained stable thanks to a multiplicity of processes that interact, interfere with, and intensify one another, thereby shaping the atmosphere that keeps our planet alive. The climate is an archetypal example of a complex system locked in a stable state.

However, since we began to industrialize, extra CO_2 has been released into the atmosphere at an unprecedented rate. In the time period of a few generations, we've burned carbon from plants that took millions of years to metamorphose into oil and gas. Current CO_2 levels are higher than at any time in the history of humanity—35 percent above preindustrial levels.[4] The last time Earth experienced levels like this was 25 million years ago during the Oligocene, when *Homo sapiens* did not exist yet and the planet was populated by animals like the *Brontotherium*, *Entelodont*, and *Proboscidea*.

"The earth was 3 to 4 degrees warmer in that era and sea levels were 75 meters higher than they are today," Schellnhuber cautions. "That's meters we're talking about, not centimeters."

As CO_2 levels continue to rise once again, the intricate web of fragile mechanisms that determines our climate will no doubt find a new equilibrium. But that new state may be wildly different from what we know today. The tipping points that Schellnhuber has identified mark the transition to that new climate. "We should try to confine global warming to a level at which we still have a chance of adapting to it. That means we ought to identify precisely which tipping points we need to avoid at any price because they will induce changes that we have no hope of handling."

URGENT RESEARCH NEEDED

Hans Joachim Schellnhuber warns that we still don't know enough about the feedback mechanisms that determine climate tipping points. "We know those tipping points are out there but not how fast they're approaching or what the consequences will be when they happen. Much closer monitoring is needed. There should be thousands of observation stations tracking things like ocean circulation and changes in monsoon patterns. We have the technology to do it, but the economic and political will is lacking. The number of meteorological observation sites has actually been declining since the 1950s."

The way tipping points interact is something else, Schellnhuber says, that we urgently need to study. The melting of the Greenland ice sheet might simply prove to be the first domino. "It will inject fresh water into the North Atlantic and increase sea levels, which might in turn slow down thermohaline circulation. That could then interfere with the West African monsoon, triggering the collapse of yet other ecosystems. The domino effect of runaway greenhouse processes has the potential to bring total global catastrophe. Astonishing as it may seem, there has been virtually no research into this kind of interaction between tipping points. Scientists have no idea how tipping in one place might impact related processes elsewhere."

To study these interactions properly, you need a computer model of the earth system that includes the atmosphere, oceans, ice sheets, biosphere, marine carbon cycle, and industrial metabolism. "We have submodels for all these elements," Schellnhuber explains, "but we need to combine them in a single, comprehensive Earth Simulator. Fitting them together is a major challenge because most of these submodels operate on different scales. But we have to manage it if we're going to be sure we're doing the right things to preserve decent life conditions for future generations. It's something we can definitely achieve within 10 years or so."

WARMING IS MASKED

Fortunately, there are certain atmospheric forces that oppose the warming of Earth. Air pollution—especially sulphur and other small particles—has a net cooling effect on our climate. Calculations suggest that if we were to pull back the "curtain" of air pollution, Earth would be more than 1.5 degrees centigrade hotter than it is today or 2.5 degrees hotter than in preindustrial times.[5]

"That's a depressing scenario after all," Schellnhuber admits. "If we were to successfully combat air pollution, we'd accelerate global warming in a way that would lead us straight into climate disaster. We would almost certainly lose the Greenland ice sheet, and we don't know whether that would in turn trigger self-amplifying dynamics. As long as we have that pollution, it will ease temperatures and mask global warming. So we shouldn't reduce air pollution too fast, as we need the cooling effect it generates. That extends our window of opportunity in which to do something about climate change."

It's tempting to dream up other atmospheric effects capable of slowing down global warming—"clean pollution" that could mask sunlight using harmless aerosols that would remain in the atmosphere for a prolonged period or by floating a film of some environmentally friendly substance on the surface of the water to reflect sunlight. There is an increasing call among scientists to resort to this means or at least to start large-scale research into the effects of it. "We could indeed escape to that kind of geo-engineering if we were absolutely desperate," Schellnhuber says. "But such procedures would be extremely risky; if a particular country stopped contributing for a while, the result could be a major climate shock. And our knowledge of the relevant feedback mechanisms isn't yet sufficiently precise. Not all air pollution has a cooling effect. Particles of carbon, for example, accelerate the warming of the earth. We need to figure out how our enemy—air pollution—can help us avoid too much global warming. We have to orchestrate our clean-air instruments in a very subtle way."

REMOVING CARBON

Hans Joachim Schellnhuber is fond of musical metaphors. He refers several times to a "symphony" of actions to counteract global warming. "We can't rely on any single instrument to reduce greenhouse gas emissions: We need a symphony of reduction measures. It will be a very complex symphony, and large parts of the score still have to be composed. But when the time comes, we're going to have to give the performance of our lives."

We could buy ourselves a considerable amount of time, Schellnhuber points out, by aggressively reducing our greenhouse gas emissions. "We have immense scope to enhance our energy efficiency. We can substitute fossil fuels for renewables like solar power. We can capture carbon from fuels and sequester it deep below the earth's surface. A symphony of actions like that could easily halve global greenhouse emissions by 2050."

The international community has started to work in this direction. An example is the Kyoto Protocol in 1997. But in all probability, halving greenhouse emissions would not be enough. "We have to understand that one degree of global warming ultimately means a 15- or 20-meter rise in sea levels. We're already past the point where we could stop that from happening unless we start actively removing carbon dioxide from the atmosphere. Even if we stopped emitting it today, the CO_2 already in the atmosphere is there to stay for millennia." It may be absorbed by photosynthesis, but not all the carbon remains fixed in plants; some of it is reemitted by the soil into the atmosphere. "We know from the carbon cycle that nature doesn't get rid of carbon dioxide entirely. A quarter of it persists in the atmosphere for a thousand years or more. So we will have to live with the climate-forcing echoes of the first Industrial Revolution for centuries to come. In the long run, the only way to keep us safe is to bring the concentration back down to pre-industrial levels. That means using biochemical processes to extract carbon from the atmosphere. Scientists and engineers need to start working on that. We still have no clue how to do it on the huge scale that would be needed without excessive energy inputs."

CHANGING SOCIETY

The scientific challenge posed by these issues is awe inspiring. "We're more or less gambling with our planet," Schellnhuber maintains. "We have to compose and perform our symphony extremely wisely. Without breakthroughs in technology, we have no chance whatsoever. But the real problem will be the responsible social implementation of that new technology. How fast can our societies absorb innovation? Perhaps not quickly enough. Will we accept the necessary constraints? These things have to be decided on a global scale; otherwise, it's a game we can't possibly win."

Little time is left to steer us away from crucial tipping points. "We have to find ways of accelerating acceptance, which may well be the decisive scientific issue now facing us. We're placing the future of humanity at risk. We have to develop communication methods and incentives capable of disseminating new technologies. Ultimately, however, you need a different type of social cohesion: We have to shape people's preferences. That's extremely delicate, but we have to do it. Sociologists, philosophers, theologians, and people of goodwill need to get together. Research is urgently required in the field of social engineering to address this problem. Very little has been done so far in that area. The challenges are frightening. But they're fascinating, too."

The alternative is equally frightening. Demand for scarce resources has too often revealed humanity's darker side. Wars have been fought over oil, and hostile migration sparked by shrinking grasslands is one of the factors currently provoking tension in Africa. If we can't prevent tipping points from being transgressed, ecosystems may collapse, and many more areas will become inhabitable. We might just find ourselves fighting each other again on a global scale. That's a terrifying prospect, as is the realization that we can only avoid it through social engineering—by basically coercing people to accept climate-friendly technology.

As we walk back down the Telegrafenberg, we take a final look at the German capital, Berlin—a city that has lived through some of the most abominable experiments in social engineering that humankind has ever devised. It is a city in which too many conductors have bellowed out their own tunes and where an entire people was persuaded to murder Europe's Jewish citizens. But it is also a city where a wall was torn down by the irresistible force of protest and prayer. A continent came together here, and history has proved in Berlin that no challenge is too great for a people that stands as one.

2.2

IMPROVING ENERGY EFFICIENCY

Imminent climate change is indeed an "inconvenient truth" that will oblige us to alter our energy use long before we begin to experience any shortage of crude oil or natural gas. Warnings about our climate come at a time when nothing would otherwise appear to stop us from using fossil energy sources for several more decades. Indeed, never before in the history of our civilization has the outlook for our continuing use of mineral oil looked so comforting. In the early 1970s, there were only 25 years' worth of known oil reserves at the consumption levels of the time. Now, at the end of the first decade of the twenty-first century, we can look forward to 42 more years of oil, even though consumption rates have almost doubled in comparison to the 1970s. Newly discovered oilfields and technologies have more than compensated for increasing demand. Known reserves are now at their highest level since we began to keep systematic statistics. That doesn't mean that progress with regard to our energy supply will be smooth; changes come in shocks, as we will see in this chapter. We are likely to witness crisis after crisis in the years ahead. And the tangible heating of our Earth will make the crises worse.

THE DYNAMICS OF OIL SUPPLY AND DEMAND

There has been a passionate worldwide debate for decades now about the precise timing of *peak oil*—the moment when oil production hits its maximum level and then begins to decline until there is no longer any left that is economically viable to extract. Yet peak oil predictions invariably prove incorrect. Each time, it turns out we can extract more oil from the ground than we previously thought. One alarming calculation after another has fallen by the wayside, as we also showed in the introduction to this book (chapter 0.2). Outbreaks of panic well up with each new wave in energy prices. Just before the credit crisis of 2008, oil prices were at a record high, and prophecies of the end of the oil era were abundant. After the crisis, oil prices collapsed, and the prophets of doom switched to preaching about the fall of international banking, the end of globalization, or the death of liberal economics.

Energy prices display very little correlation with real production costs or proven reserves. Shifts over time reflect the discovery of oilfields and technological progress in response to our increasing demand for energy. Although no new oil is being created inside the earth, more reserves every year come within reach of our technology. New wells are discovered at regular intervals, and new extraction techniques continue to be developed. Modern drilling platforms can tackle much larger fields and can drill much deeper than their equivalents 10 years ago. They can drill at an angle, too, enabling them to extract oil from several sources at once. These techniques can in turn be applied to smaller fields. In the past 20 years, this has led to a 30 percent increase in the amount of oil that can be extracted around the world. The outlook has thus evolved from one year to another.[1]

It is interesting to take a closer look at the pattern of these changes. The outlook for our oil supply remained more or less flat in the 1990s compared to the sharp fluctuations the graph shows for the 1970s and early 1980s. It turns out that larger and smaller fluctuations alternate in a special way. The graph has much in common with the patterns displayed by avalanches and earthquakes. The magnitudes of all these events obey scaling laws. That's a discomforting thought, as it suggests there is no equilibrium between the forces regulating supply and demand. Our oil markets are characterized instead by tensions that build up, reinforce one another, and accumulate to a point where a small shift may trigger a landslide. Substantial fluctuations in the availability of oil are more likely than we would wish. Because the oil economy is nowhere near equilibrium, major shifts are probable. That means we should expect more crises and sudden price rises but also the emergence of unforeseen new reserves and energy-related technologies that relieve the tension. Our energy economy is a self-organizing critical system that uses shocks to adapt to changes.

And those changes will come whether we prepare for them or not. If we sit back and wait, these changes will come as a shock, possibly unleashing catastrophic upheavals. To avoid that, we need to prepare.

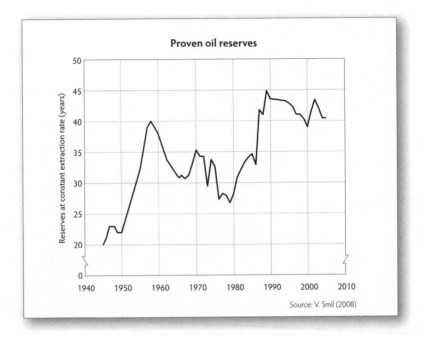

The outlook for our continuing use of mineral oil is more reassuring now than it has ever been. Advances in oil technology continue to outstrip the increase in consumption. It is clear from the chart that the discovery and adoption of new reserves tends to be associated with a series of shocks. That's a discomforting thought, as it suggests there is no equilibrium between the forces regulating supply and demand. And those shocks could increase in severity as we get nearer to depleting our oil reserves. More efficient use of the reserves could help us mitigate them to some extent. *Source*: Smil, V. (2008). *Energy in nature and society*. Cambridge, Mass.: MIT Press.

We can soften shocks through cleaner technologies that can be readily slotted into our rigid infrastructures, making them more responsive to change. These technologies may not be able in themselves to prevent irreversible climate change or the eventual depletion of our mineral resources. But they can at least give us a little breathing space so that we can pursue a more permanent fix.

In the remainder of this chapter, we describe a number of transitional technologies and strategies that could make energy systems more flexible and responsive to varying circumstances. We will come back to the more fundamental changes needed to achieve a low-carbon energy economy in the next chapter (chapter 2.3).

THE ISSUES BEHIND SAVING ENERGY

An obvious way to gain time is to use energy more efficiently, the potential for which is amazingly large. There are substantial world-wide differences in energy efficiency in nearly everything we produce and use. The energy intensity of the Japanese economy is only half that of the United States. For every dollar produced, Americans use double the amount of energy that the Japanese require. This indicates that most countries have plenty of scope for improvement. Televisions, refrigerators, and PCs could all use a lot less energy. If we were all to perform as well as the best available commercial technology, worldwide energy consumption could be halved.[2] And it wouldn't require new technology. We can start to reduce our energy consumption immediately. In most cases, that will save money as well.

Take the streamlining of cars. Designers currently have little incentive to reduce the air resistance of their vehicles, which means that cars use an unnecessary amount of energy. Even models that look sleek could still be improved substantially. By making a few simple adjustments, Greenpeace managed to reduce the Renault Twingo's air resistance by 30 percent.[3] Techniques like this aren't applied, however, because designers and buyers are more interested in other things. Prominent wing mirrors sell better than streamlined ones. The same goes for almost everything we use. The technology is here already, and in many cases, it's profitable, too. As consumers and producers plainly don't care about these savings, efficiency should be implemented through regulations and campaigns. In a democratic society, this obviously leads to a tension between private will and public interest.

Energy conservation has another drawback, however. Higher efficiency doesn't necessarily mean lower total energy consumption in the long run. On the contrary, raising efficiency almost inevitably results in higher consumption. The reasons for this counterintuitive effect of efficiency gains are simple: Any money we save is invariably used for extra consumption. Homes have become more economical to heat and keep cool, but we now live in bigger houses with fewer people. Aircraft are more efficient, but we fly more often. Televisions use less power, but we now have sets in the bedroom, kitchen, and bathroom as well as the living room. We also buy televisions with much larger screens. Saving energy does initially have an effect, therefore, but this is subsequently canceled out by extra consumption. That means the improvement of efficiency must be a continuous process, with one measure constantly following another, necessitating steady investment in technologies.

It is interesting to compare this with energy use in nature. Species didn't evolve to maximum efficiency in energy use. They evolved to maximum fitness, as Charles Darwin put it. In 1922, U.S. physical chemist Alfred Lotka tried to translate this fitness in energy terms and came up with a principle that was later controversially coined the Fourth Law of Thermodynamics.[4] Those organisms survive that manage to use a maximum of power, Lotka stated. Or expressed differently, organisms tend to maximize the rate of useful energy use through them. Energy efficiency is balanced against the need for high energy density. The latter determines power and performance. The Lotka principle is valid for a broad range of energy-consuming systems, not only in nature. It is also valid for products, buildings, transport and other economic activities. The fittest technologies maximize power density. Energy efficiency is still valuable, as it increases the amount of energy that is available for useful purposes. The efficiency of energy conversion, however, is constrained by the need to have maximum energy flow. This often goes at the expense of optimizing a system for the largest overall energy efficiency. So there is an intrinsic reason, caused by competition, that overall energy is inefficient. A high throughput is more important for survival, but it costs energy.

For chemical processes, in industry and in living beings, a high yield is essential. This means that the amount of material produced per unit time and space is optimized rather than the energy consumption. A chemical reaction that takes place at high speed is often far from equilibrium and thus less efficient than a slow reaction close to equilibrium. It was calculated that in many practical situations, an energy efficiency above 50 percent would not serve a useful purpose in the struggle to survive. Increasing efficiency further would limit the throughput and reduce the flexibility to switch to another energy source if necessary. An extreme energy conservation ultimately causes a lock-in in one energy technology. This reminds us of the ideas of Thomas Homer-Dixon (chapter 0.2). He claims that an increased efficiency causes an increased sensitivity to shocks when circumstances change. This decreases fitness in the Darwinian sense and reduces adaptability. Modern technology allows for conservation measures that also make our energy systems more flexible and adaptable to varying circumstances. In the following section, we consider the example of fuel for transport.

FUELING OUR TRANSPORTATION

What came first, the car or the service station? A car is worthless if you can't fill it up. But nobody is going to build a service station if there aren't any cars passing by. Cars and gasoline evolved together over more than a century to the point where the infrastructure now forms a rigid system that can't be easily changed. That's what makes our refineries such a technological marvel. Generations of chemists have devoted their creativity to the plants' gradual improvement. Anyone thinking of opening a biodiesel plant will have their work cut out for them: They will start developing their technology from the point petroleum chemists had reached 50 years ago. Anyone for solar cars? They will face a similar dilemma.

Whereas new technologies must start from scratch, existing infrastructure can be easily expanded. New refineries are often built alongside old ones, with access to the same harbors. New oil tankers virtually roll off a production line, all following the same established design principles that made their predecessors so successful. Even in the case of an expressway, it's easier to widen an existing route than to build a new road. The layout has a greater longevity than its components. If you aim to introduce greater efficiency and flexibility, therefore, the most promising strategy is to change the components rather than the system itself. It is a promising development, for instance, that some cars run on both gasoline and electricity because the infrastructure for both is already in place. Their combination saves energy because the car's electrical generators and batteries ensure that the gasoline motor always operates at optimum efficiency. It also offers flexibility, as hybrid cars are less dependent on a single energy infrastructure. That makes them easy to introduce. Hydrogen cars have a much tougher future because there is no hydrogen infrastructure in place.

Likewise, it is difficult to replace our refineries with other fuel-production plants. Instead, existing refineries can be adapted to take more than one feedstock, which would make them less dependent on a single mineral energy source. Or, one step further, we could incorporate technologies to make liquid fuels from natural gas and coal. This would also make refineries more flexible, while simultaneously saving energy and reducing carbon emissions, as combined refineries like this are better able to adapt to changing demands. They can be designed to take biological material as feedstock, too. Biofuels could partially replace liquid fuels, which would also reduce carbon emissions. It's generally a bad idea to grow crops specifically for use in fuel, as this invariably competes with food production and water provision. But straw and waste wood can also be used. Even more promising is the use of algae, which offer ten times the energy of a crop of soybeans per unit of surface. That's because land vegetation expends most of its energy in pumping up water from the soil and outgrowing its neighbors, whereas algae can use every spark of energy to procreate.

Sustainable biofuels like this could fuel our aircraft, where switching to electricity is not an option.[5] They would make aviation less dependent on mineral oil, giving us time to work on alternative long-distance transportation networks, such as high-speed rail and fast ferries. At the same time, we can work on the many technological breakthroughs that are necessary to bring closer the use of hydrogen as a transportation fuel (see the next chapter).

THE RIGIDITY OF OUR ELECTRICITY GRID

The electricity grid is another rigid element within our energy infrastructure. Its configuration constitutes a major obstacle to change, according to Jan Blom, Professor Emeritus of electrical power systems at Eindhoven University of Technology in the Netherlands. Blom is also a former director of the Dutch electric power research and regulatory organization KEMA. Thus, he knows the electricity supply infrastructure inside out. "The more bulk power lines you build, the more rigid the network becomes. That not only means that blackouts are more likely; it also becomes more difficult to carry out changes. Altering the route even of a single power cable can lead to serious problems. In densely populated areas, high-voltage pylons can't be rerouted because there is no space for new ones. We have to make the grid less rigid so that we can respond to change more rapidly. We need to make our power grid and generators more flexible for the future."

Our existing power grids are geared toward a centralized approach for the generation of electricity. You can spot a power station from a long way off by the high-voltage cables that fan out from it in every direction, distributing the plant's output to consumers. The current centralization of electricity generation is technical in origin. For many years, bigger meant more efficient and cheaper. A larger boiler, for instance, was less susceptible to heat loss. "Increased scale resulted in savings," Blom explains. "But the technology is now so good that size is no longer important. We can make smaller and smaller turbines, for instance, without sacrificing efficiency. That means we can configure the grid in a different, less centralized way, with generation closer to the user. That will reverse the current hierarchy: Rather than having a central supply with passive customers, we'll move toward a situation in which customers themselves generate energy—with a home-heating system, for example, that also generates electricity. Households will then be able to supply energy back to the network." That's already happening, albeit on a small scale. "The network's central control rooms wouldn't be able to cope if there was a significant increase in the number of power suppliers. Control is the real challenge in terms of decentralizing our electricity supply," Blom warns.

The current control strategy is to assume that power only flows in one direction. Sensors within the network are largely confined to the backbone. When all the power flows from the center, operators don't need to know what's going on in the network's capillaries. They have a general idea about the dynamics of the network based on the changes in voltage and frequency that they measure. The frequency provides long-distance information, and the voltage tells them about the status of the network at shorter range. If your neighbor switches on the dishwasher, for instance, you might notice a slight voltage dip. These signals make some central control possible without an additional computer network for control.

The model is changing as more and more consumers start to generate electricity for themselves. This makes it necessary to have more detailed insight into the behavior of the networks, something that global parameters like voltage and frequency can't provide. The first step toward more dynamic control would be to increase the number of sensors in the local branches of the networks. More decentralized intelligence should follow. "The goal is to design self-regulating electricity grids in which thousands of smaller units all decide for themselves. So-called agent technology will enable this," Jan Blom says. "Decentralization means a reversal of the grid hierarchy. It will make the electricity supply more flexible and allows us to include many more small generators in the grid. The key question is how much extra communication is necessary for this decentralized technology. It would be wonderful if you didn't need an extra, separate control network, as that would hamper the reliability. So the local agents should ideally be able to function on the basis of the voltage and frequency they measure at their location within the network. But we still don't know if that is possible."

A further step in reversing the current hierarchy would be for consumers to become smarter in their energy use, Blom continues. "The control of the network has always been based on the concept that supply follows demand. If the load changes, the amount of generation is adapted accordingly. However, if more variable sources like wind energy and solar energy are included in the generation of electricity, it makes sense to shift the concept and have demand follow supply. That would mean, for example, that the dishwasher starts up as soon as the wind begins to deliver enough power. That would enable us to exploit our renewables to the maximum extent."

Decentralization automatically means that connections over larger distances will become less important, Blom says. "So we shouldn't keep making power lines thicker and thicker. Instead, new lines should be built to increase redundancy in the network. If each location is accessible via multiple routes, you have more possibilities for rerouting power as demand or supply changes." Energy storage is also important when decentralizing the grid. If electricity can be stored close to consumers, local energy shortages can be supplemented locally, reducing dependence on long-distance connections. Battery technology is evolving rapidly, so this is not a remote option. "The introduction of electric cars in particular will spur this storage option," Jan Blom thinks. "The battery in your car could also be used to store electricity in the grid at night."

We could eventually have a self-organizing grid that automatically adapts when new generators are plugged in. "That would allow new technologies to be incorporated in the infrastructure much faster. It's a genuine revolution that will take us from centralized generation to *power to the people*," Blom says. The idea reflects developments in other areas, too. The Internet likewise serves numerous providers and users at the same time. Meanwhile, self-healing, rerouting, redundancy, and decentralization are issues that are pursued in this other key infrastructure for our society as well. Making our electricity grids smarter is only a first step toward making our critical infrastructures more flexible. We also need to prepare for more fundamental changes in our energy supply. In the long run, other energy sources will have to take over from crude oil. That's the subject of the next chapter.

2.3

SEARCHING FOR NEW ENERGY

We looked in the previous chapter at the prospects for our current energy infrastructure and asked how we can make it more flexible and sustainable. In this chapter, we fast-forward to the new energy economy after the oil era. The quantity of available energy is not the main worry in the postcarbon era. We're surrounded by tremendous amounts of energy. The power of the sun's rays was there long before we started to discover fossil energy sources, and on Earth's surface, we can harness wind and water. Another vast amount of energy is encapsulated in our planet in the form of heat. As yet, we only tap small fractions of these natural energy supplies. Evaluating our long-term options, we have to ask ourselves: How can we harness these energy sources in such a way that they may serve us without a serious regress in our human civilization? Only then may we hope for a gradual transition to a new energy era.

In the course of our history, we have used ever more concentrated forms of energy. In the era when we warmed ourselves by a wood fire and ate the grains of the field, we needed about 1 square meter of land for each watt of energy that came available. When we tamed wind and water power, the energy yield of a square meter of land rose by a factor of ten. The advent of coal, oil, and gas accounted for another factor of hundred improvement. This is calculated by summing up the amount of land you need for excavating the energy carriers and converting them to a useful form of energy. A similar calculus can be made using the energy content of the energy carriers themselves. Society has evolved with each subsequent energy innovation. More concentrated forms of energy allowed for a more concentrated community with a more complex division of labor. Now we don't have to search large areas of land for some useful calories for ourselves; we can devote our time to comfort and complicated products.[1] That's another formulation of the evolutionary principle of Alfred Lotka that we encountered in the last chapter. A society seems to evolve toward a maximum power density.[2]

So the energy issue is not only about digging up fossils or catching the rays of the sun. It's also about making it available for use, which means that we have to *concentrate, transport,* and *store* it. That's what we'll call the *energy triad.* All three legs of the triangle are important in funneling the energy into our complex society. A future society needs energy in dense forms that may be transported and stored to sustain its elaborate structures. Failing to fuel the complex human fabric may cause a decline. Thomas Homer-Dixon has argued that the Roman Empire started to decline when it had exhausted its best energy resources, which in that time took the form of croplands.[3] For

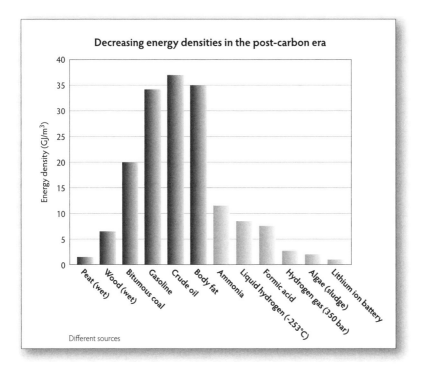

Up to now, our civilization has evolved in parallel with the increasing energy density of our fuels. The fuels listed here for the postcarbon era don't offer the necessary increase in energy density, and neither do alternative sources like the sun and wind. To continue to fuel our megacities and high-speed infrastructures, we must either seek higher-density sources of energy or else develop conversion technologies that increase energy density.

its food-based energy system, it had to move to poorer lands, its energy supply lines snaked farther and father from its major cities, and the population had to work harder and harder to feed the oxen that were instrumental in building the complex structures of Roman society. So in the energy triad, defects arose in the concentration, the transport, and the storage of energy. A similar breakdown has been described for other societies. Arguably, the world dominance of the Netherlands in the seventeenth century was based on peat and wind energy, which came to an end when other countries started to use more concentrated fuels, something the Netherlands lacked at that time.

Finding an alternative for our current fossil-based energy economy requires a system approach. We not only must take all three legs of the energy triad into account, but we also must reckon with changes in the structure of societies. So society and politics will need to be involved in preparing for the transition.

ENERGY FROM THE SUN

So let's start our analysis with solar energy. In theory, the sun offers us more than enough energy to power our entire civilization. The solar energy received by Earth's surface is 10,000 times what we consume. Two hours' worth of solar radiation would be enough to provide all of humanity with energy for a year. But the sun's rays are spread over a large area, significantly reducing their intensity. On average, every square meter of Earth's surface receives 170 watts,[4] from which even today's best solar cell technology can extract only a maximum of 30 watts—barely enough to power a lightbulb. Modern photovoltaic cells perform better than wind turbines per unit of surface but lag far below the concentrated forms of fossil energy that we have grown accustomed to. The roof area of a standard home in the countryside may be big enough for all the solar cells that would be needed to meet the building's energy requirement, provided that the building is extremely well isolated and the efficiency of the appliances is the best possible. But it is certainly not enough for the majority of humankind that lives in old houses or high-rise buildings. Our age needs more concentrated forms of energy. A city can by no means be self-sufficient when a significant proportion of its power consumption needs to be generated by the sun. That's the reason the present use of solar energy is so marginal. It's easy to garment your roof with solar cells, but that would power only a small fraction of a household's electricity. Everything beyond that would require totally different infrastructures to concentrate the energy, transport it, and store it. We need to consider the complete energy triad to make solar energy viable. Not only do we have to develop solar cells further, but we also must create new infrastructures to bring them to our cars, offices, and homes.

In a real solar economy, an immense area would have to be covered with solar cells, which means that it would require us to transport electricity over large distances. We would have to install long cables from distant deserts and create major east–west connections to carry energy to different time zones. Plans like this face the problem of high costs, drawn-out procedures, and political complications in selecting a route. The Eumena supergrid, for example, based on high-voltage transmission of African solar energy, would have to run through countries like Libya, Algeria, and Tunisia; that would create new dependencies.[5]

The entities in the energy triad are communicating vessels. When we would like to diminish the need for long-range links, we have to raise the efficiency of local generation and storage of solar energy so that less energy needs to come from elsewhere. Electricity doesn't seem the optimal medium in two of the three sides of this triad. Transport of electricity over longer distances is cumbersome, and it introduces stability issues, as described in the previous chapter. A continental or even global network of long-distance power connections would increase sensitivity to shocks within electricity systems. Electricity can't be stored very well either. This is a serious shortcoming in view of the intermittent character of solar radiation. The sun has often set by the time people turn on their televisions in the evening. Batteries and water reservoirs are by no way sufficient to smooth out variations in daylight and energy use. Breakthroughs in these fields are clearly necessary.

Another significant challenge is the large-scale production of solar cells. Most technologies become cheaper with mass production, as miniaturization reduces material costs and the price of machines. That happened, for example, with central-heating systems. Sometimes, such mass production is reached by an increase of scale, which also makes technology cheaper. That's long been the secret of success for power plants. In the case of solar cells, however, neither strategy works. If you miniaturize the surface area of solar cells, you capture less light. And if you scale the technology up, you simply need more materials. There is no such thing as Moore's law for solar cells. The same goes for many other forms of renewable energy, such as windmills or hydropower. The yield is directly proportional to the surface area of the solar cell, the amount of material used for a windmill, or the area of the water basin.

Raw materials are a bottleneck for solar cells, as producers can't turn them out fast enough. Since 2007, more silicon has been used in the production of solar cells than in that of microelectronics, rapidly leading to shortages of the precious silicon wafers. The shortage has taken on such acute proportions that some silicon manufacturers now refuse to sign contracts for periods of fewer than 10 years. The companies that build the machines required to produce solar cells can't keep up with demand either. As a result, anyone wishing to set up a solar cell factory has to wait for up to several years to get hold of the necessary equipment and materials. If we were to shift to a really significant level of solar power, we would have to find a technology that is more easily scaled up with materials that are readily available in large amounts and machines that don't need operators with 5 years of university education. One breakthrough might be the use of conducting polymers to produce lightweight and flexible solar cells. These can already be printed in roll-to-roll processes at speeds that exceed the throughput of any other solar technology by far. Although state-of-the-art polymer solar cells currently achieve power conversion efficiencies of about 6 percent, the technology is still in its infancy. There are promising ideas for increasing that efficiency.

WINDMILLS

Many of the drawbacks of solar energy also apply to the large-scale use of wind power. All three legs of the energy triangle need to be strengthened for many of the same reasons as is the case with solar energy. Wind is a variable and dilute form of energy that requires large surfaces, long-range networks, and mass storage if its use is to increase substantially. Unlike solar power, wind energy is very much a mature technology. Tremendous advances have been made since

Dutch engineers began to use this technology in the seventeenth century to drain their low-lying country and create new habitable land. Today's windmills can harvest 50 percent of the energy carried by the moving air, which is very close to the theoretical maximum of 59 per cent.[6] The drive to improve the technology has thus been successful. The latest models don't have to be shut down in high winds, and flexible materials and designs have improved performance under a wide range of conditions. The improvement in reliability and maintenance is equally impressive. Taken together, these advances have brought the price of wind energy within range of the electricity generated by gas-fired power plants. Windmills have truly come of age. There's not much room for further refinement. So we can't expect a substantial decrease in cost per unit of power. Yet the yield is still less than half the output of modern photovoltaic cells when deployed over the same area of land.[7] If solar cells continue to improve, they will outperform windmills.

Despite its impressive growth, wind power still constitutes only a small percentage of any single country's electricity supply. Sparsely populated and windy Denmark holds the world record, with 20 percent of its electricity generated by the wind. That's about as high as you can get. It is increasingly difficult to find suitable locations for new installations, which means extra growth will have to be achieved by shifting to less favorable locations. We simply can't imagine a wind economy in which more than half our industrial power is provided by windmills, as was the case in the Netherlands in the seventeenth and eighteenth centuries. Wind power is, as we said, a mature technology with the potential to deliver only a modest proportion of our total energy requirement.

HYDROPOWER

The hunt for suitable sites is also an issue for hydroelectric power. The necessary reservoirs require large tracts of land, and building hydroelectric dams often displaces local populations. Between 40 and 80 million people were obliged to relocate in this way in the twentieth century. And if you compare the surface area of these reservoirs with the power they help to generate, hydroelectricity frequently performs less efficiently than tapping the sun's rays directly. Many dams achieve only a tenth of the wattage per square meter offered by solar cells. But there are major differences between individual hydro-electric plants. The Three Gorges Dam currently being constructed in China will require less space per unit of output than solar cells do. Hydroelectric reservoirs are also less environmentally friendly than you might think because of the gases they give off. Rotting plants often release considerable volumes of methane—a more powerful greenhouse gas than carbon dioxide—into the atmosphere. As a result, some dams contribute just as much to global warming as conventional power stations of a similar output. Another drawback is that capacity declines over time because a great deal of sediment is left behind in the reservoir. And the immense weight of all that water can affect the geological balance in unpredictable ways, increasing the likelihood of earthquakes.

Therefore, not everybody views hydroelectric power as a renewable source of energy. Things were very different in the 1960s and 1970s, when thousands of dams were built—500 a year throughout those two decades. In the United States in recent years, more dams have been dismantled than have been constructed, which is easier said than done, as they were seemingly intended to last forever. The builders of several dams failed to consider that the reservoirs would have to be emptied once they came to the end of their useful life. It's a common pattern when it comes to our energy supply: The possibility that ideas and needs will change is simply ignored. The reservoirs of hydroelectric plants are like bathtubs without a plug. Consequently, holes now must be drilled laboriously through the thick concrete walls of several decommissioned dams.

One of the great benefits of hydroelectric power is that it has a very strong storage leg in the energy triangle. Hydropower can help smooth out fluctuations in energy consumption. A full reservoir can be used to generate electricity whenever you like because a turbine can be activated in a matter of seconds in response to prevailing demand. For that reason, water power is ideal for backing up the output of wind turbines and solar cells. If the wind dies down or clouds weaken the sun's rays, the power company can increase hydroelectric output, which means hydroelectricity could potentially contribute to the growth of other green energy sources, too.

In an energy economy that relies primarily on solar energy, water reservoirs don't store enough power to level off fluctuations. There are simply not enough mountains to extend this function much further. Little further expansion would be possible because the most attractive locations have already been taken in most regions with the exception of Africa.

Dams, however, have another important function. They can help mitigate fluctuations in water consumption, which could make them doubly attractive in African countries, where several promising locations are available (see chapter 1.1).

GEOTHERMAL ENERGY

One source of sustainable energy has been systematically overlooked, maybe because of the devastating power it unleashes whenever it comes to the surface. Volcanoes spew out immense amounts of energy in a short space of time and with tremendous violence. There are several fortunate places around the world where Earth's natural heat is conveyed upward in a much less disruptive manner in the shape of hot water. In the Philippines, for instance, a quarter of the country's electricity is generated from geothermal energy. Iceland exports the energy it derives from its famous geysers. This kind of heat can be found underground elsewhere, too. To exploit it using the technology available today, you'd have to create artificial geysers. Several approaches have been suggested for "enhanced geothermal systems" of this kind. In one design, cold water is pumped down a specially drilled shaft until it boils; it is then returned to the surface in the form of steam via another shaft. These shafts must be really deep, though, penetrating up to 10 kilometers below the surface of the earth.

Geothermal energy would slot perfectly into our present energy infrastructure. A considerable amount of energy can be derived from a single source, making this an attractive way to generate electricity.

The construction of artificial geysers has been tested in France, Australia, and Switzerland. The largest example is operated commercially at Innamincka in South Australia. It has already generated its first power and will produce 50 megawatts by 2012—enough for a town of 50,000 inhabitants. Geothermal energy is extremely promising; the interior of the earth is an energy resource on a scale capable of servicing our entire civilization for millennia to come.[8] But there are still major challenges to overcome. The injection and heating of water risk triggering earthquakes—something that has already happened in Switzerland. We also lack experience when it comes to drilling hot impermeable rocks like granite, which oil companies have traditionally avoided because they don't contain any oil.

In the energy triad, the legs of storage and transport are readily available for geothermal energy. That is, the energy source is in fact an enormous reservoir from which heat can be extracted as it is needed. Unlike solar and wind power, it is available round the clock, in all seasons, and could be readily regulated. And the distribution fits seamlessly in our centralized power infrastructure. However, the leg of generation, or in this case extraction, is not mature yet. But this could change. A breakthrough in drilling technology and technologies that can transform heat into consumable energy could make large amounts of energy available in the next 20 years. This energy source doesn't appear in long-range energy scenarios because it is not economical to use with our present technology. But radical cost reductions are not uncommon for a novel technology. What's more, scaling the technology up would further decrease costs, as many of the tools for drilling holes can be reused. A geothermal economy is not less promising, therefore, than a solar economy.

NUCLEAR POWER

Like geothermal power, nuclear energy has the advantage that it dovetails well with our current centralized ways of generating and distributing electricity. Power is produced on the same large scale as in conventional plants, which means increased use of nuclear energy wouldn't oblige us to reorganize the grid. And nuclear power doesn't contribute to global warming either. However, the pace at which new power plants are being commissioned is too slow to keep up with the current growth in energy consumption. To double the number of megawatts generated by nuclear energy in the next 20 years, we would need about 1,000 new power plants. This means that we have to build one plant a week. Reducing carbon emissions to any meaningful extent will require at least a tripling of nuclear power. Increasing the share of nuclear power would be extremely expensive. The safe reactor technologies developed in the aftermath of Chernobyl and Three Mile Island have made this form of power generation even more costly than it was already.

To fuel a carbon-free economy in the long term we should also switch to nuclear fuels that are more abundant than the uranium-235 used in most reactors. There is a reactor type for which the depletion of uranium is less of a problem. Breeder reactors make their own fuel, which means they operate with a closed fuel cycle. These reactors are powered by the much more common uranium-238. Unfortunately, this technology is also perfect for making nuclear bombs. Breeder reactors are more complex than conventional reactor types as well, making them even more expensive than other nuclear plants. Fear of nuclear arms proliferation combined with the prohibitive costs has kept this technology from getting off the ground. We take a closer look at the proliferation issue in chapter 5.6. For now, it suffices to remark that the construction of a Dutch–German breeder plant in Kalkar, Germany, sparked such vehement protest that it was never actually commissioned. The concrete structure was eventually converted into a theme park called Nuclear Water Wonderland (*Kernwasserwunderland*). One key advantage of breeder reactors is frequently overlooked—namely, they can be designed in a way that minimizes the output of nuclear waste. This justifies a further development of breeder reactors. But to be useful on a large scale, several breakthroughs will be needed in reactor safety. Hence, the generation leg of the energy triad requires substantial research. In addition, the storage leg needs attention, as nuclear power plants are often difficult to operate with a variable load.

ENERGY CARRIERS

We have to work on all these new technologies to make a transition to another energy era possible. The lead time for this kind of technology is pretty long, so we need to start right away. And we'll have to work on the storage and distribution of that energy as well. That's particularly clear when we think about the energy use in cars, trains, and aircraft. If we want to make our transport carbon-free, we'll need to find alternatives for gasoline, diesel, kerosene, and jet fuel. That's the subject of the last part of this chapter. Important breakthroughs are necessary here because few alternatives are available. In the previous chapter, we mentioned fuels derived from biological materials such as algae as a transitional solution. Eventually, only small quantities of biofuel can be produced in a sustainable way, as we argued. There are few other possibilities. The most obvious is the use of electricity, as most of the innovative forms of energy that we have mentioned in this chapter generate electricity. But electricity is extremely hard to store on a large scale. The storage leg in the electricity triad is virtually lacking. The energy density of a battery lags far behind that of gasoline. And the price per stored unit of power is still extremely high. We may be able to store enough power in batteries for electric cars, but large ships and aircraft will continue to require some other kind of portable energy carrier for several more decades at least.

The other possibility is to synthesize fuels. Hydrogen is the most obvious candidate, as it can easily be produced using electricity, just as in secondary school electrolysis experiments. When an electrical current is passed through water, oxygen and hydrogen are given off. But electrolysis is not very efficient: 30–40 percent of the energy is lost in the process. The technology involved is expensive, too. Electrolysis systems need to become much cheaper.

The benefit is that hydrogen is much easier to store than electricity. It closes that side of the energy triad. It's extremely light, too: 1 kilogram of hydrogen contains almost three times as much energy as the equivalent amount of gasoline. Liquid hydrogen was chosen to fuel the Space Shuttle precisely because of this considerable weight saving. Cars can benefit in the same way. Hydrogen is especially useful for road vehicles when combined with fuel cells. In a highly efficient process, fuel cells generate electricity from hydrogen, which is then used to power an electric motor. Fuel cells are more efficient than internal combustion engines. But before that, a major advance will be needed to make fuel cells cheaper; we have to get them down to about a tenth of their current price. Part of the high cost is due to the fact that the cells require platinum. The scientific challenge is to develop new techniques that make fuel cells that are cheap and can be produced on a really large scale. But we still have a long way to go. A technical breakthrough is also required in terms of hydrogen storage. Hydrogen may be light, but it takes up too much space. Hydrogen is a rarefied gas, which means it has to be kept under high pressure to squeeze a sufficient amount of it into a small tank.

Long term, hydrogen holds out the prospect of truly clean transport. Significant breakthroughs will be necessary, however, before it can genuinely bring about greener mobility. The main challenges lie in the efficient production and compact storage of hydrogen. This is an area where we have to explore really new ideas that require long-term research. New ideas can be found at the forefront of chemical research.

We turned to Johannes Lercher, professor of chemical technology at Munich University of Technology. He proposes a novel technology that turns coal into hydrogen. Coal is found in many more locations than oil is, he notes: "There's enough fossil carbon to last another 600 to 1,000 years. So we need to develop the technology to turn coal into a sustainable energy source. The problem with coal is that it yields less energy per unit of carbon than either oil or natural gas. So at first sight, it's not exactly an energy source that can serve us in the postcarbon era. Energy companies are developing techniques for capturing the CO_2 and pumping it back into the ground, but that is an expensive and energy-intensive process."

Lercher proposes a revolutionary technique for processing coal while it is still underground by building a kind of chemical plant deep within the seam itself. "The idea is to gasify the coal in situ, creating hydrogen and carbon dioxide, which you then separate using a filter. You bring the hydrogen to the surface, and you leave the CO_2 underground." Oil companies are already experimenting with underground techniques but only in oilfields. The principle is somewhat more difficult in a coal seam. First, you have to make the coal porous so that you can gasify it underground. One way of doing that, Lercher suggests, might be to use ultrasound. Another problem is the high temperature needed for conversion. An underground coal seam is generally less than 100 degrees, whereas gasification requires temperatures in excess of 300 degrees. "You can get that temperature down using catalysts. If we can develop the right ones, catalysts could lead to a breakthrough in this technology. We could then convert coal into hydrogen underground, which would be fantastic." Catalysts are substances that facilitate a chemical reaction without themselves undergoing any change. They enable us to carry out processes at a lower temperature or more convenient pressure. And deep underground, they might help us produce hydrogen.[9]

This option is obviously still far off. But it promises a higher efficiency for hydrogen production than electrolysis can offer. An option for the storage of hydrogen is suggested by the research of Danish colleagues of Johannes Lercher. At the University of Lyngby, an idea has been developed to store hydrogen.[10] As long as it is difficult to store hydrogen in a small volume, and hence, to take it with you, its use will remain limited. Hydrogen must become much denser before it can flow with some chance of success through our societies. One way of making hydrogen denser is incorporating it in larger molecules. Ammonia (NH_3), for example, can be stored in much denser form than hydrogen. The hydrogen content is three times as high as hydrogen that is stored under the very high pressure of 300 bar. Making ammonia from nitrogen in Earth's atmosphere is one of the oldest, most widespread, and best optimized processes in the chemical industry; these processes now produce 100 billion kilograms of fertilizer a year. The hydrogen is released as soon as ammonia is decomposed again. That would make hydrogen available in its pure form so that it can be burned or used in a fuel cell to generate electricity. The resultant nitrogen may be safely released back into the atmosphere, or it can be reused in a new cycle of energy storage. The efficiency of this cycle may be approximately 80 percent.

As ammonia is very noxious, it's not suitable for transport. But the idea of incorporating hydrogen in larger molecules can be taken broader. Hydrogen can also be combined with carbon dioxide. Chemists in the German town of Rostock are pursuing this idea.[11] It is more complex than the use of nitrogen because we still have no technology to extract carbon dioxide directly from the air with sufficient efficiency. The difficulty is its low concentration. It constitutes only 0.038 percent of the air in our atmosphere, whereas nitrogen constitutes 78 percent. But it is possible to extract carbon dioxide from the exhausts of power plants. This is now done at some places for carbon capture and storage (CCS) to reduce the emission of carbon dioxide. But instead of storing it underground in empty gas reservoirs, the captured carbon can be given a second life as energy carrier. In a later stage of development, carbon dioxide may be extracted from the air.

Hydrogen and carbon dioxide can be converted into formic acid (HCOOH). This reaction has long been studied, and efficient procedures are available. It produces a liquid that can easily be stored and transported. It contains twice as much hydrogen as compressed hydrogen at 350 bar pressure. It is less flammable than gasoline or ethanol. And it's not noxious; in small quantities, it is even allowed as a food additive in the United States. To release the hydrogen again, HCOOH can be decomposed into hydrogen and carbon dioxide.

One step further would be the conversion of hydrogen with carbon monoxide to gasoline-like hydrocarbons. That would close the carbon cycle. By burning fuel, we get carbon dioxide, which can in turn be converted into fuel. Carbon dioxide extracted from air is eventually returned to air. It would cost energy, of course, but this can be generated with solar cells, geothermal sources, or other sustainable techniques. The synthetic fuels would close the energy triad, offering a possibility for storage and transport.

The process to make synthetic fuels was invented in Germany in the 1920s to produce an ersatz fuel based on gasification of coal, which meant Germany no longer had to import petroleum. Nowadays, it is used on a significant scale to produce liquid fuels from natural gas or biomass. The challenge is to reach the same level of selectivity, efficiency, and flexibility that we have accomplished in a century of development of fossil-based petrochemistry.

There is still a lot of development necessary before these ideas can be put into large-scale practice. But many elements are already available. Surely, the use of hydrogen isn't an option that we can implement tomorrow. In the previous chapter, we highlighted the difficulty that the transition to an entirely new fuel like hydrogen would pose for our transport systems. It might, however, be a possibility for the long term, provided that the necessary breakthroughs can be achieved. However, 20 years from now, renewable power will still not be the dominant element within our energy supply. The difficulties in upscaling new energy technologies and closing the energy triads are too big for that. Without closing the energy triads, the transition to another energy era will not be gradual but an uneven process driven by external stimuli. Society will have to pay constant attention to these new energy sources. There is also much technological development needed to achieve all this. The introduction of new technologies will require an enormous effort in creativity and capital. This will require bold actions of politicians. For now, there is enough energy, and climate change hasn't yet hit hard enough to make action inevitable. Everywhere in the world, new coal-fired power plants are still being built. That is terrible proof of the indifference of politics.

If we can succeed in reversing that indifference, there is plenty of scope for new growth. In our view, the greatest long-term potential may be offered by geothermal energy in combination with solar and supplemented by a modest amount of wind power. New chemical processes can then close the energy triad. If we don't succeed, our grandchildren will have reasons to worry. Energy is the basis of everything we do—not just energy for production or transportation but also as feedstock for the chemical processes that produce everything from plastic to pharmaceuticals. The continuous supply of energy is one of the greatest challenges humanity faces.

2.4

SUSTAINABLE MATERIALS

Volkswagen's new car plant is located more or less in the center of Dresden, Germany. Through its glass shell, you can watch the fitters from the street as they work at the production line. On another storey, robots spray the car bodies in bright colors without a single splash of paint winding up on any of the windows. Back in the twentieth century, it was customary to push factories beyond the city limits because they were too dirty to share our living environment. Today, clean, attractive, and compact factories are finding their way back. Production techniques have reached such a state of perfection that we're eager for them to be seen. Factories and houses can once again stand side by side. A hundred kilometers away in Leipzig, BMW has also made a symbol of this kind of modern, clean manufacturing. The production line at the company's new plant runs straight through the cafeteria. Cars float above the tables as lunch is served below. The automotive industry's production techniques are every bit as clean nowadays as those of the firms providing the food on the workers' plates.

Cars are only one example, but they are an icon of our industry. Every last detail of their production process has indeed been fine-tuned. Yet the cleanness of its production is only superficial. It hasn't resulted in the perfect car. Far from it: Motor vehicles continue to impose an ever-heavier burden on the environment. It takes more raw materials to produce one now than it did 20 years ago, and cars still use the same amount of fuel for every kilometer they travel. That simply can't continue if we still want to visit friends on the other side of the country two decades from now. Breakthroughs are needed that will make cars more energy efficient and environmentally friendly. And the same goes for other modes of transport, such as trains, aircraft, and ships. Apart from further inventions, we're going to need entirely new materials, construction methods, and production techniques if the boundaries are truly to shift. The question is whether our industry still has the necessary flexibility to achieve any such breakthrough.

A CAR IS A CAR IS A CAR

Designers have invested thousands of hours in every component of the modern car, perfecting the form, choice of materials, and manufacture of even the tiniest part. In so doing, however, they have lost sight of the bigger picture. Ton Peijs, professor of materials at Queen Mary University of London, finds that incomprehensible. He cites the example of painting techniques: In the latest car factories, robots spray the final layer of paint—the shiny one—with extreme precision. The car is electrically charged so that the paint is applied evenly to the bodywork. "It's been optimized down to the last detail, and the result is a car that positively gleams. It could be done differently with much lower cost and less environmental impact, but that would mean changing the entire production process." Body panels for cars could be made out of thermoplastics with which the desired color could then be mixed. But you can't introduce a change like that by simply adjusting one element of the production process. The entire assembly sequence would have to be altered, requiring the kind of fundamental reorganization that can only be achieved by building completely new factories. That has, in fact, already occurred: German carmaker Daimler built a brand-new plant to manufacture its Smart model, which features a large number of plastic body panels. It's a substantial investment, especially when you're not certain whether consumers will accept a car that looks so different. "Using precolored plastic panels results in a matte finish we're not used to seeing in cars," Peijs explains. "It's also extremely difficult to match the color of the panels with that of the metal components, which is why Smart cars deliberately use plastic elements of a totally different color."

Car manufacturers nowadays construct different types of cars on the same production lines, but that doesn't mean that changes can be made easily. On the contrary, it poses constraints that are anything but flexible. The layout of the new BMW plant in Leipzig illustrates how rigidly production is organized. The different models made on its production line are built in precisely the same way, in the same order, even if the components are different. Each car gets its color, its dashboard, its wiring, and its seats at the same place, and that will also be the case 10 years from now, even though different models will no doubt be rolling off the line by then. An engineer who would like to cast the wiring into the chassis has no chance of ever seeing that done, any more than a designer has who wants to manufacture the chassis and body as a single unit. Changes like that would change the sequence of operations, which automatically rules them out. The principles of manufacture are generally fixed before the engineers even start to think about the next model of car.

Even a brand-new factory tends to adopt ideas wholesale from its precursors with few fundamental changes. Surprisingly little has changed since Henry Ford set up his first production line. New types of engines hardly get looked at because gasoline and diesel designs have been evolving for more than a century now. Imagine what a Stirling engine might look like today if the entire world had devoted a similar amount of time to improving it.

Engineers only deviate from their manuals when strictly necessary. Experience is passed on and continually developed, making it progressively more difficult to take an alternative path. This lack of flexibility acts as a drag on the industry's evolution. The way a car is manufactured has already been fixed for the decade ahead. Only the individual components can still be altered, and so that is what car designers focus on.

DIVERSITY ISN'T ALWAYS GOOD

This focus on individual components means that the car has become a jumble of materials and techniques. Much of the steel has been replaced by plastic, which is lighter and in many cases cheaper and easier to manufacture. Ton Peijs points out that a different plastic is often chosen for each component: "Cars frequently include components made of twenty-five or more different types of plastic, which makes them difficult to recycle. We are making progress in this field, as this number tends to be limited in newer cars. Still, steel cars were far better in that respect. You had a single material that could be easily melted down and reused. We need to reduce that diversity further; otherwise, recycling will remain too complicated."

So far, only limited success has been achieved in that regard. Audi and Jaguar are trying to develop cars that are substantially constructed from aluminum, but that particular metal is expensive, and it dents easily. BMW and Mercedes are working on a car that uses only a few different types of plastic. Even then, frequent exceptions turn out to be necessary, such as reinforcing the plastic with fiberglass in certain places. The resultant material is harder rather than easier to recycle. For its part, Toyota is pinning its hopes on biodegradable plastics and composites, which Peijs considers equally naive: "It doesn't close the material cycle. You lose the amount of energy that is contained in the plastics. And why would you want to compost a car at the end of its life? No one has a clue what we would do with that amount of biodegradable material. It's full of additives that no farmer will like to have on his land. You have to think these things through properly; otherwise, there's no point."

The variety of materials in cars is steadily increasing. Experts are developing ever more specialized synthetics, Peijs explains: "They're working on synthetics that are transparent, conductive, fire resistant, strong or light, or which offer a combination of those properties. The trend is toward smart additives, which you use in small amounts to precisely adjust the material's properties according to the specific component you want to make. But that means you can't recycle them anymore. It is possible, however, to alter the physical characteristics of plastics without changing their chemical composition. That's a better solution." Peijs mentions a type of polypropylene developed at the beginning of the century. It is reinforced with fibers that are themselves made of polypropylene. Another possibility is to produce fibers that are thinner—around a nanometer across. In that case, you need fewer additives to achieve the same strength. Reducing the amount of added fiber also makes it easier to recycle.

LIGHTER MATERIALS MEAN HEAVIER CARS

The rise of plastics doesn't mean that vehicles have become any lighter. Although components and materials have been improved in thousands of ways over the years, designers tend to seize on every improvement as an opportunity to add even more extra features. It's always easier to do that than it is to alter the entire concept of a car.

As a result, the performance of the car as a whole didn't improve. The pattern is most obvious when we compare different generations of the same model. Some cars are produced over several decades, allowing us to track their evolution from one generation to another. Metal parts have become thinner over time as design methods became smarter, and many models have experienced the gradual replacement of metal with plastics. New features include soundproofing, electric seat control, and power steering. Safety was increased by reinforcing the bodywork and by adding seat belts, headrests, and air bags. Weighty climatization adds to that. As cars became heavier, they needed better brakes and more powerful engines, which also weigh more. Cars in Europe have been gaining weight at an average rate of 4 kilograms a year. Over a 20-year period, the Toyota Corolla and the Vauxhall Cavalier/Opel Vectra have gained more than 40 percent in weight,

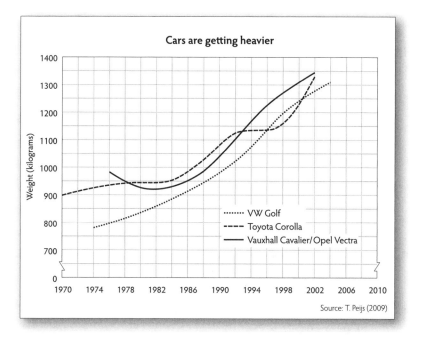

Cars are made of lighter materials than they were a few decades ago, but that hasn't stopped them from getting heavier. Efficiency gains have been systematically used to make cars safer and more comfortable with the result that fuel consumption has remained roughly the same. The challenge now is to rethink car design in a way that improves the functionality and at the same time reduce fuel consumption. *Source*: Peijs, T. (2009). Personal communication.

and Volkswagens have gained as much as 50 percent. The power requirement has risen accordingly. A car that doubles in weight uses almost twice the amount of fuel, wiping out all the gains of more efficient engine technology.[1] That's why cars still use the same amount of fuel per kilometer as they did 10 years ago despite all the technical advances in engine technology. Total energy consumption has actually risen: There are more cars, and they all clock up more kilometers per year.

"We have to break that trend toward steadily heavier cars," Peijs insists. "It would be a real breakthrough if we could produce lighter and simpler cars in the next 20 years, reversing the spiral of increasing weight. Lighter bodywork means you can also have lighter axles, wheels, and engines. You could actually halve the weight of the car. Less weight means lower fuel consumption, better acceleration, and higher speeds. On a global scale, it would save an immense amount of materials and energy and would also result in better cars."

NATURE HAS LEARNED TO BE VERY SMART

Lots of engineers have been turning to biology as a potential source of change. "It's extremely inspiring to look at nature," Ton Peijs confirms. "Nature is smart." Emulating the natural world has become something of a craze among engineers, who are desperate to find out how caterpillars crawl, how trees bend, how insects fly, and how geckos grip. Nature has found elegant solutions for all sorts of technical problems with only the strongest, lightest, most sustainable, and economical forms winning the battle for survival. These solutions—developed over millions of years—are invariably the best and are environmentally friendly to boot. Anything inferior has died out along the way. So why reinvent the wheel?

Peijs points to seashells as an example. Only 5 percent of a shell consists of polymers, with the remainder formed around them from chalk and other minerals. Together, the result is 3,000 times stronger than regular chalk. "We're nowhere near that; it's extremely difficult to copy," Peijs admits. "Shellfish, algae, and trees frequently produce better materials than we're capable of. That's why we need to learn from nature."

Nature sells, too. Mercedes designers advertise the fact that they drew inspiration from the trunkfish, and BMW has used sharkskin to market its improved streamlining. Opel, meanwhile, highlights parallels between the way trees grow and the engine suspension system it developed for its Astra range. Continental promotes a car tire inspired by the legs of that master climber, the tree frog. The parallels with nature are often a little superficial, but they plainly make for good marketing.

SLAVISH IMITATION

"Nature's smart solutions are sometimes imitated rather unintelligently," Ton Peijs admits. "Engineers often focus too much on mimicking nature rather than on studying its underlying principles and methods and then applying them to their own designs. But that's what you need to do if you're going to take ideas from nature and use them in a different context. It's about more than just slavish imitation." Nature doesn't work in the same way as engineers, who prefer to think in terms of physical properties—choosing materials according to their strength, flexibility, or resistance to wear. They like to use high-performance materials to produce high-quality products, whereas natural materials often have much humbler properties. Nature, you might say, focuses on smart design techniques rather than on high performance. It takes a weak material and makes it rigid, for instance, by adding reinforcement in precisely the right spots. "Shape is cheaper than material," as Peijs puts it. "That's an important natural principle. Nature uses materials that we would call inferior and moulds them in the right structure to give it superior properties. There's more to be gained through smart design than there is by developing better materials."

Complexity is another principle of nature. Peijs cites the example of bones: "Everything is built in. They're self-healing, they have embedded sensors, and they're constructed with nanoprecision using fiber reinforcement to deliver extra strength precisely where it's needed. You can organize their structure into seven hierarchical levels, each with different mechanisms that determine the bone's properties at that particular scale. All the elements are closely interrelated. It would be a major breakthrough if we could emulate that. Our materials have a maximum of three levels of scale and are frequently optimized in terms of a single property such as rigidity or strength."

Nature doesn't rely on a single recipe to produce its structures. Instead, it can call on an almost unimaginable variety of solutions. Nature tries out alternative designs by means of random mutations. As we saw, human beings have standardized manufacturing methods to such an extent that cars are now produced in exactly the same way, no matter what they look like or what specific characteristics they are given. Nature teaches us that we should make the logistics in a car factory less rigid so that it is easier to try out different design concepts.

"Nature is economical, too," Ton Peijs adds. "If it comes up with a lighter construction, it does so precisely in order to make an organism lighter—enabling it to move faster and use less energy. Whenever we come up with a lighter material for car manufacture, by contrast, we seize on it as an opportunity to pack the vehicle with even more luxuries." On the downside, nature can be slow, which is perhaps the greatest challenge facing designers who look to it for inspiration. It takes months to weave the precise structure of a seashell. A car factory can't afford to wait that long. "We need to take what we see in nature and then reproduce it at a faster rate. Otherwise, it's not worth our while. The great challenge," Ton Peijs concludes, "is to speed up that slow, natural process of creation."

Nature continues to outperform us. It produces materials and uses construction concepts that are superior to our own. That means we've not exhausted the potential for improvement. But we're going to have to tap that potential more effectively if we're serious about reversing the ever-increasing burden we're placing on our environment.

2.5

CLEAN FACTORIES

If you ask a child to draw a factory, you'll most likely see a picture of huge chimneys pouring out dark smoke. Adults might come up with associations like explosions, barren industrial estates, and wasted energy and raw materials. Chemical plants, with their endless pipes and weird smells, have a particularly bad name when it comes to damaging our soil and atmosphere. Chemistry today is—we have to admit—far from ideal. Many of the industry's perceived sins relate directly to its gigantic size. You only have to look at our power stations or the factories that produce our plastics: They're growing bigger all the time. They often need huge cooling installations to get rid of all the excess heat. This is just another way of saying that they use far too much energy.

Bigger plants bring bigger dangers. Things can go very badly in a large installation. The repercussions of an accident can be dramatic, which is why safety is such a key feature when designing them. But that imposes restrictions on the installation's operations. It often means that we have to operate the processes far from the optimum. Operators need to play it safe at the expense of additional material and energy consumption. Truckloads of by-products must be removed—often in such vast quantities that there's hardly any useful purpose they can serve. In classic refining techniques, for instance, it's hard to adjust the ratio between light and heavy oil products. If you need a lot of gasoline, you end up with an excess of fuel oil, or vice versa.

Increased scale has long been the chemical industry's watchword and for compelling technical and financial reasons. A large vessel, for instance, is easier to insulate than a small one. There are other arguments in favor of large scale: Investment costs, personnel levels, maintenance, administrative costs, and land use have all traditionally been lower per unit of product in a big plant. Until recently, it's always been an issue of bigger meaning more efficient and cheaper.

Nowadays, however, the classic approach is incresingly unnecessary. The gains to be achieved through upscaling are steadily diminishing, and the logistics associated with ever-larger plants are mind-boggling. The drawbacks of scale are begining to bite. Ethylene crackers, for instance, are so big these days that the construction of a single installation immediately disrupts the global balance of supply and demand. Competitors need to market their products jointly so as to distribute capacity evenly and avoid excessively large jumps in output. That's one of the reasons for the ethylene pipelines that link production locations all over Europe.

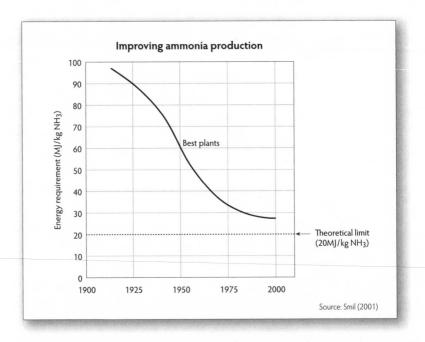

Decades of technological development have brought more efficient and cheaper chemical processes, such as here for the production of ammonia, one of the most used processes on the globe. Increased scale has brought many of these improvements. However, the drawbacks of scale are beginning to bite. The challenge is now to decrease the size of a chemical plant without sacrificing efficiency. *Source*: Smil, V. (2001). *Enriching the earth: Fritz Haber, Carl Bosch and the transformation of world food production*. Cambridge, Mass.: MIT Press.

SIZE MATTERS

Technical progress means that the advantages of large plants are no longer so critical. Better insulation techniques, for example, can also benefit installations with a smaller output. Thanks to advances in electronics and process automation, sensors and operating systems are not only better but cheaper, too, making them attractive for small installations as well. It also means that production at smaller plants no longer automatically requires more personnel. So it is perfectly possible to decrease the size of a chemical plant without sacrificing efficiency. Small vessels have the additional advantage that you can bring them to the right temperature faster, as you don't have to heat up such a gigantic volume. There's greater precision, too; you need to apply heat only at the spot where the chemical reaction is occurring. And even when the reaction is violent, you can control it more readily on a small scale. These benefits tend to become greater when the plant is smaller. So why not reverse the size spiral and shrink the equipment? The entire panoply of chemical installations can be made much smaller. You can miniaturize all the vessels, pipes, and distillation columns that make up a chemical plant—down to the size of your hand or even smaller.

Throughout the world, researchers are working diligently to create microplants. The idea is to use the same microscopic etching techniques that we find in microelectronics. It is now possible to construct devices to a precision of a few micrometers. This opens up all sorts of new opportunities for downscaling chemical processes. Microstructures that are normally used to isolate one electronic component from another lend themselves perfectly to carrying fluids. Sensors may also be patterned on silicon. It's even possible to etch small moving structures onto a chip. Micropumps and micromixers are, therefore, already a reality. The relevant technology was significantly boosted by the needs of DNA research. In the 1990s, technologists successfully integrated all the techniques required for gene analysis into a single microchip. The resultant "labs on a chip" have enabled us to automate the many complex experiments needed to sequence the complete human genome (see chapter 4.2).

Klavs Jensen is one of the pioneering chemists who is taking this idea a step further. So far, the approach has only been applied to laboratory equipment; the next breakthrough will be to use it in chemical production to create a plant on the scale of a chip. Jensen, who is professor of chemical engineering and materials science at the Massachusetts Institute of Technology, has set himself the ultimate goal of integrating all the stages of an industrial process in a set of interconnected devices much like a computer with its display, central processing unit, and storage devices. All the familiar processes at a large chemical plant—mixing, reaction, and separation—can be performed within a "factory on a chip" of this kind. Jensen says such an approach might represent the ideal chemical plant: "If you make them really small, you can control chemical reactions drip by drip. You'll be able to introduce exactly the right amount of raw materials. You can then remove the reaction products immediately without them reacting further and forming by-products. The final product will be of a higher quality and will be more sustainable because less material and energy will be used." That's very different from a large reaction vessel into which you dump everything at once and where the reaction products are frequently left behind until the very end.[1]

MANY DROPS MAKE A RIVER

Microplants are much safer than the large industrial installations to which we're accustomed, Jensen continues. "Even if a microreactor were to fail, only a small quantity of chemicals would be released. It would be easy to contain, and the defective unit could be readily replaced." In addition to the safety aspect, the small scale makes it easier to perform reactions under optimum conditions. "More aggressive conditions can be used. Exothermic reactions tend to 'run away,' triggering a chain reaction that ultimately leads to an explosion. That's always an issue, for instance, with oxidation. Reactions like that have to be handled with extreme care. Industrial oxidation processes are designed very conservatively. A microreactor, by contrast, can be operated closer to the point of runaway. In a large-scale installation, we'd never dare to do it."

Excess energy can also be put to better use on the small scale of a microplant. Endothermic and exothermic reactions can be combined, with the energy released in one reaction fed into another. Another advantage of microplants is the greater uniformity of fluids in the microchip. "That's because it's easier for us to measure and control a process locally. We can achieve greater precision, which means you end up with fewer by-products and can avoid incomplete reactions. Increased precision also enables you to perform reactions that can't be done on a larger scale. It creates the possibility of new reactions that are too difficult for conventional equipment." The processes needed to produce materials like Teflon are a good example. Teflon is a solid material with unique inertness, resistance to most chemicals, and thermoresistance. But the fluorination process used to manufacture it is highly corrosive and exothermic. The small scale offered by microplants would make that process easier to control.

Jensen also cites the crystallization of complex fluids by pharmaceutical producers. The precise form of these crystals is crucial to their activity in the human body, which means they have to be very precisely synthesized. Minor fluctuations in pressure or temperature frequently result in crystals of an entirely different shape. Microreactors offer the perfect tool with which to ensure the most effective crystal shape. They allow for systematic variation of crystal forms and enable new pharmaceutical agents to be designed and tested rapidly. This has already become a widespread technology used to automate trial and error in drug discovery.

The future for miniature factories like this is all the brighter when we consider that their small scale also creates fresh strategies for rolling out industrial processes. In classic chemical production, a new procedure is first tried out cautiously at the laboratory level and then gradually scaled up one step at a time until industrial production is achieved. In a microplant, you can skip all those steps. To increase capacity, you simply raise the number of microplants. This can be done gradually by having more and more devices operating in parallel. "Rather than building larger and larger structures, it will be a question of more of the same. That will make it easier to respond rapidly and flexibly to new developments in the market or to consumer demand," Klavs Jensen believes. And it won't be necessary to concentrate all those parallel units in one place: "You'll be able to install microplants wherever their products are required. Which also means you won't have to transport toxic intermediate products like cyanides. You can manufacture them precisely where they're needed." Much the same is true for products with short lifetimes. Jensen gives the example of the contrast fluids used in medical equipment like PET scanners, which are normally only usable for a few hours. "Using microplants, you could produce them alongside the PET equipment," Jensen says.[2]

Microplants could even be built into domestic appliances. Washing machines could produce their own detergent, and soda siphons could make their own soft drinks. Microplants could also be sited at the same location as the raw materials. This would create plants capable of, say, refining oil while it is still in the well. Biomass could be processed near the fields in which it is grown. "The future lies in smaller scale and more precise production closer to where it is needed." The reduced scale of microplants also creates the possibility of an entirely different approach to logistics. We currently have centralized manufacturing, which pushes us toward uniform production and is vulnerable to disruption. Microplants will enable us to produce on a small scale, close to the end user, and with the ability to adjust rapidly to changes in the market or to new requirements.

OVERSEE IT ALL

The real challenge in terms of microplants lies in bringing together the different parts of a plant in a single design. The traditional procedure is to begin by researching the processes and then to design a reactor. In many cases, control is little more than an afterthought. When it comes to a microfactory, by contrast, everything needs to be created in one go. The reactor *is* the process. Sensors and electronics will have to be tightly integrated in the resultant microplant, which makes producing chips and microreactors a truly multidisciplinary task. "It means bringing in new knowledge from the outside," Klavs Jensen confirms. "I've brought biologists and chemists into my group. You need them. You have to learn from one another." Jensen's team is working on the bottlenecks that hinder progress with microplants. "We want to achieve truly three-dimensional structures on a chip. The vessels, pipes, and columns that make up a classic chemical plant create a sophisticated knot. The layout of a large plant is optimized to keep transportation to a minimum. So far, it hasn't been possible to overlay elements within a microplant, which limits the potential for optimizing layout." Processing solid materials and crystallization are also difficult because solid particles can block the tiny channels.

The first microplant applications will be in fine chemicals and the pharmaceutical industry—two sectors in which precision is crucial and where output is frequently measured in small quantities. In pharmaceutical production especially, precision provides an immediate payoff. The accuracy offered by microplants might also lead to better products when synthesizing plastics. The precise length of the synthetic molecules could be controlled, allowing the fine-tuning of their properties. Klavs Jensen is convinced, however, that microplants have much to offer in terms of bulk processes as well, in which output is measured in tons rather than grams. Oxidation processes can be performed in a microplant under such favorable conditions that it will pay to put large numbers of them alongside each other. The study of microplants will also offer fresh insights into how to integrate processes, which will in turn make for better designs on a larger scale. This, Jensen believes, will enable us to control complicated processes more effectively.

This will bring us closer to plants in which everything is totally controlled and coordinated with no by-products, excess, or wasted energy—a plant that automatically adjusts to the conditions in which it's operating and that is located close to the user. Microplants might ultimately allow us to create a "universal factory" in the shape of highly versatile devices based on the integration of different components on a single chip. A variety of products could then be made using just one unit. One day, we might even be able to manipulate molecular subunits at will, forming molecules in much the same way as a printer produces different types of text.

Part 3
TOOLS

3.0

OUR ASSISTANTS

Canadian media guru Marshall McLuhan predicted the rise of the "global village" back in 1962.[1] Time and space, he said, would cease to be barriers to communication, enabling people to form relationships on a worldwide basis. In the past 10 years, rapid growth in communication opportunities has validated much of his analysis. All the same, the world has not turned into one great village. Whole regions of our planet have been excluded, as we can see from the map of the world's Internet connections. The major links bypass the continent of Africa. From the Atlantic Ocean, they touch the Cape of Good Hope before arcing onward to the Pacific, with just the occasional minor branch to the African coast. They look much like the trade routes of the old Dutch and English East India Companies, in fact. A cable running through Africa would be far too vulnerable, even assuming that any local people or businesses could afford fast Internet connections in the first place. So it is that an entire continent can miss out on the communication revolution, causing it in turn to be shunned by the business world. Software firms develop their programs in China and India rather than in Cameroon.

A denser network of communications could give people a greater opportunity to participate in the global economy. It might also give them more control over their water supplies or provide them with early signals of global change. Many other problems that humans face are technical in nature, as are the tools we need to confront them. Microelectronics offers tools to better monitor our health. And more flexible, error-aware computers could steer us away from crises. We need tools that are responsive and ubiquitous. We need to measure and control larger areas on a shorter timescale and with much greater accuracy than is currently possible. We still don't have enough sensors to monitor our climate or imminent earthquakes. We consume too much energy and too many raw materials in our manufacturing plants because we don't know how to control the processes more accurately. New tools could allow us to keep our finger on the pulse and respond quickly if things threaten to go wrong. Influencing the fate of humanity is not simply a question of the big picture and long-range forecasts; it's also about remaining alert to small changes.

Attempts to improve our tools have always sparked anxiety. Human history is full of warnings that new technology will do us irreparable harm and ultimately enslave us. Socrates cautioned against writing, and it was also argued that the invention of printing would change us beyond recognition. And it probably did, too, though not in the way feared by late-medieval ecclesiastical leaders. Our own age also displays deep-seated fears of humankind's tools taking over. American molecular scientist Eric Drexler, for instance, has postulated that molecular systems could reproduce, eventually creating a new type of self-replicating nanorobot.[2] This replication, he fears, could occur so rapidly if sufficient raw materials were available that swarms of nanorobots would soon be spilling out of every hole, crack, and pore, devouring the matter around them like single-celled organisms and churning out fresh copies of themselves at lightning speed. Thousands—and before long, billions—of nanorobots would spread across Earth voraciously consuming everything in their path. Drexler's supporters calculated that the entire planet could be stripped bare in this way in less than 3 hours, turning everything into "gray goo." Within 180 minutes of a nanorobot escaping from the laboratory, all life on Earth would be wiped out.

It's a nightmarish image that has cropped up in all sorts of variants. Bill Joy, the former head of research at computer maker Sun Microsystems, has offered a slightly more subtle version in which nanorobots achieve humanlike intelligence.[3] Ray Kurzweil, the American futurologist, has speculated about the effects nanorobots could have on our bodies.[4] They could swim through our veins and take over our internal functions. They could fire electrons at the right moment in our brains, triggering our neurons with artificial signals. Whatever our eyes were actually seeing, we might believe ourselves to be on a tropical island watching the sunset. We'd hear the birds singing while in reality we were in a factory operating heavy machinery. Nanorobots would fool our senses, making it impossible for us to distinguish reality from illusion. The idea is taken to its logical conclusion in the movie *The Matrix*. Human beings will ultimately find ourselves being exploited without us even noticing.

People like Drexler, Joy, and Kurzweil express an unease that resonates strongly in our society. Exaggerated fears about the tools we develop are just as baseless as claims that we're heading for some kind of technological utopia. Both distract us from the genuine problems we face. We plainly need substantial progress in science and technology to survive and prosper. We should not be discouraged by the dangerous side of technology. Every successful new development has a potentially negative aspect. Knives have been used to kill humans as well as to save many others in hospital operating theaters. Helicopters can be lethal weapons, but they are often the only vehicles that can reach remote villages hit by an earthquake. We shouldn't fret about science and technology, therefore, so much as the society using them.

In this part of the book, we first describe how electronics and computer networks reach their limits and how new breakthroughs may enlarge their scope (chapters 3.1 and 3.2). Then we go into the dilemmas of allocation (chapter 3.3). Important tools for protecting privacy and property come from cryptography (chapter 3.4). Managing imperfection is an important theme in the creation of software and hardware (chapter 3.5), in logistics (chapter 3.6), as well in the control of robots (chapter 3.7).

3.1

SMARTER ELECTRONICS

It can be pretty dispiriting to go into an electronics shop shortly after you've bought yourself a new computer. Every month, you seem to get more computing power for the same money. And every year, the devices get smaller. The pace of progress is so intense that you're out of date again within a few months. If you look back 10 years, the most advanced computers of the time now appear clumsy. Ten years from today, our current computers will no doubt seem ridiculously primitive, too. It may be frustrating, but that same rapid progress offers an important source of hope for humanity, as it will enable us to apply the power of computers to areas that have so far proved resistant. All sorts of pressing issues in health care and the global economy stand to benefit from the ever-decreasing price of electronics and the ability to pack more and more computing and communication power into a smaller space. That's why electronics lies at the heart of many of the solutions we describe in the remainder of this book.

The promise held out by electronics amounts to more than the wishful thinking of unworldly nerds laboring in dust-free labs to develop yet more powerful microchips. It is bound up with the nature of the new kind of problems we face. Electronics is proving increasingly important in complex situations that are difficult to control. A tiny oscillation in the earth's interior, for instance, has the potential to induce an earthquake. It's therefore vital that we install inexpensive devices around the globe that can pick up these early signals before they intensify, while there's still time to warn the people in harm's way. The same applies to many other complex contemporary issues. Powerful computers could routinely analyze masses of financial data to detect the next crisis in the making, and a few years from now, we may have computers powerful enough to identify tipping points in the climate. Rapid calculation, accurate measurement, and automatic control will give us a firmer grip on complex problems.

Cheap, ubiquitous, and powerful electronics will play a crucial part in improving the human condition. So it's a very good thing that components are getting steadily cheaper and more powerful. Gordon Moore—cofounder of chipmaker Intel—foresaw this incredible increase in computing power and the lowering of its price. Back in 1965, he predicted that the number of transistors on a chip would double every 2 years. His forecast turned out to be on the conservative side: The number of components on a microprocessor has actually doubled every 18 months or so. Moore's law, as it came to be known, has held up unshakeably for 40 years now.[1]

One reason the miniaturization of electronic components has proceeded in such predictable steps is the simultaneous evolution of the tools used to design them. The first computers were sketched out using pen and paper and were built with wires and screwdrivers. Chip designers then used those first computers to perform the calculations needed to design subsequent processors, and since then, each new generation of computers has delivered the tools to produce the next. Designers now routinely use fast computers and intelligent software to conceive chips at a high level of abstraction. The latest microprocessors, with their billions of components, couldn't possibly have been shaped by human beings directly. The components on a chip are now designed by computers with virtually no human intervention. So new computers make new processors possible, and vice versa. These and other factors have made Moore's law an economic and microelectronic truism: The experts draw up "roadmaps"[2] in which Moore's exponential line is a self-evident goal factored into all planned developments in the field. In this way, it has become a self-fulfilling prophecy, as engineers set out, roadmap in hand, to halve the size of their components over successive 18-month cycles.

THE LIMITS ARE IN SIGHT

Many people in the chip industry are convinced that Moore's law will hold up for a few more generations. However, the physics of ever-shrinking devices isn't susceptible to wishful thinking or challenging roadmaps. How sure are we that Moore's law will continue to deliver steadily cheaper electronics for the next 20 years? Can we really develop affordable devices capable of empowering all the world's people? To find out, we turned to Hugo De Man, who has devoted his professional life to creating ever-faster and smaller electronics. He began his career as a young boy building radio receivers with vacuum tubes before graduating to transistors as a teenager. As professor of design methodology for complex ICs in Leuven, Belgium, he learned how to pack more and more transistors onto microelectronic chips.

Hugo De Man cofounded IMEC, one of the handful of places around the world that have kept Moore's law ticking. Manufacturers like Intel, Samsung, and NXP have joined forces at this world-renowned research facility to develop new microelectronic technology. De Man, who has recently retired from Leuven's University and IMEC, believes we are witnessing the final stages of the miniaturization process. "In the '80s and '90s, we enjoyed a prolonged period of 'happy scaling,' as we call it here. Generation by generation, we made everything progressively smaller while keeping the transistor structure and its materials the same. It was also around then that Intel declared that you could continue driving up the clock speed of a single processor up to 10 or 20 gigahertz." Things suddenly became a lot more difficult around the turn of the millennium, at which point details measuring 100 nanometer became possible. At that degree of miniaturization, De Man explains, "We're approaching lots of different limits at the same time. The problems are stacking up, making it extremely difficult to take the next steps."

First of all, one must realize that all computational activity generates heat. The faster the processor calculates, the more heat is produced. Today's microprocessors are dissipating as much as a 100-watt lightbulb in a few cubic centimeters. This heat output doubles at each new generation. Removing that heat by air cooling becomes a challenge. One way to escape this is by lowering the voltage at which the processor operates. Half the voltage means four times less heat. That sounds great, but unfortunately it also reduces the speed of computation. As a result, the speed at which single processor chips operate can't be raised much further. "The speeds Intel was talking about a while ago are now fantasy. We've reached the point where even a slight increase in frequency causes a single processor chip to generate substantially more heat. That makes it virtually impossible to produce reliable chips with a frequency of more than 5 gigahertz: They'd just melt."[3]

A number of solutions have nevertheless been developed to allow processor chips to perform more calculations per second. Instead of using faster transistors, we can achieve speed gains, for instance, by integrating multiple cores on the same chip that operate in parallel and allow you to carry out several calculations at once. These processors can then operate at a low voltage to solve the heat issue. That delivers increased speed but not for every computer program. "You actually need to redesign most software to make optimum use of parallel configurations," De Man says. "But that would be practically impossible. The software we have right now was built up gradually over decades. You can't redo it all at a stroke. Instead, all the big manufacturers are working on ways of automatically adapting software for parallel computing, although no one has come up with anything yet. In practice, there's no point in placing more than eight processors in parallel for traditional general-purpose programmable applications. Adding more doesn't give you any extra speed. We've now more or less reached that limit. However things are different for specialized applications such as image processing, computer graphics, and speech processing. These kinds of programs have a lot of intrinsic parallelism and can effectively be implemented on application-specific chips containing hundreds of small processors operating in parallel and generating less heat. This is also crucial for the battery-operated mobile and wireless smart phones and netbooks that gradually take over from the PC."

Another way to overcome the speed issue is to improve the speed of the classical transistor, for example, by the incorporation of new materials. However, here we are hitting another physical boundary because such transistors cannot be switched off properly. As a result, they continuously leak energy even when not calculating. Design engineers invented leakage-suppression techniques, but these tricks also gradually become exhausted. The only remaining solution is to start making transistors based on different physical mechanisms—tunnel field-effect transistors (TFETs), for instance. But that would mean breaking with 40 years of microelectronic design techniques. It implies bidding farewell to the classic transistors the young Hugo De Man used when building a radio and which we've since learned to cram into ever-smaller spaces. Switching to TFETs would require totally different design tools, for instance—tools that don't yet exist.

Another issue arises when trying to produce features below 32 nanometers as are needed for the next generation. Today, this is done by lithography using ultraviolet (UV) laser light of 193 nanometers. "Amazing things have been achieved," De Man says. "We've learned how to use light to create structures almost ten times smaller than the wavelength of the UV light patterns. Everyone thought that would conflict with the laws of physics, but we developed a comprehensive box of tricks that enabled us to do it. In 2003, we were producing structures measuring 90 nanometers; by 2009, we'd gotten it down to 32 nanometers, all using the same laser light wavelength of 193 nanometers." But the bag of tricks to etch patterns onto a chip is now almost empty. Existing ultraviolet lasers are probably good for one more generation. Designers came up with the clever technique of using water as an extra lens, but that's as far as it goes. Further shrinkage will only be possible using laser sources of extreme ultraviolet (EUV) light of 13.5 nanometers wavelength. IMEC's labs is now testing the first EUV prototype machine in the world, fabricated by ASML in the Netherlands, capable of manufacturing chips with details smaller than 22 nanometers wide. Such machines are very expensive because no more lenses can be used at that wavelength. Extremely precise mirrors operating under a vacuum must focus the EUV light. The mirrors need to be extremely flat, varying no more than the equivalent of a millimeter every 1,500 kilometers.

Another issue is the inevitable inaccuracy that creeps in as components continue to decrease in size. The performance of extremely small transistor structures starts to depend on only a few hundred atoms that are, by nature, randomly spread in silicon. As a result, transistors that are supposed to be identical display tiny variations, causing them to behave differently and in some cases to malfunction. Dealing with such random variations in component properties on a chip will require entirely different design methods (see chapter 3.5). Fault-tolerant design will force software engineers to take the details of the hardware into account, Hugo De Man says. "Software makers have traditionally viewed the processor as a black box, which they don't need to know anything about. That's going to change. Software and hardware design will inevitably become intertwined. It will be impossible to separate the tasks of the hardware builder from those of the software designer, all of which will make the design process a lot more complicated."

It's the first time in the 40-year history of microelectronics that so many challenges have coincided, De Man confirms. "There's a cumulative set of problems. To keep up with Moore's law, we'll have to solve them all in the next few years. The essence of the law is that you have to solve problems in an ever-shorter time frame. It's particularly clear if you plot Moore's law on a linear scale; that's when you realize the explosive increase in performance we've achieved time after time. It's taken the whole of human history to arrive at the processors we have today. We now have to repeat that achievement in just 4 years, and then in 2 years, then just 1, then 6 months, and then 3. That's Moore's law. The pace of improvement has to rise all the time if you want to keep on that exponential line."

MOORE'S ANTILAW

"Let's say we can surmount the current challenges," De Man says. "I doubt if that will be possible at the usual pace, but for argument's sake, let's say it can be done. What would we achieve? It would mean that in 10 years' time, we'll be able to make a processor on a single square centimetre of silicon that can perform 25 trillion calculations per second (25 TOps). Or to put it another way, that can handle 150 different HDTV channels simultaneously. Do we really need that? Are we really willing to pay for it?"

What does each new generation of chips actually cost? "Today, it takes about a $100 million to develop the hardware and software for a new processor," according to De Man. "In 10 years, the cost is more like $1 billion. The price doubles every 18 months—Moore's antilaw, I call it. You hear a lot less about it than you do about Moore's law, but the explosive increase in costs is putting a brake on progress as the production volume for adequate return on investment runs in the hundred of millions of chips. The same goes for a lot of the techniques you need for microprocessors. The costs of setting up a new electronics plant are also growing exponentially. Do you really want to spend $10 billion on a new fab? The R&D cost for developing the next generation process technology is reaching $1.2 billion for the 22-nanometer node."

Historically, rising costs have mostly been addressed by sharing these huge R&D costs: IMEC—the institute that Hugo De Man cofounded and where several different players have pooled facilities and know-how—is a reflection of that process. At the

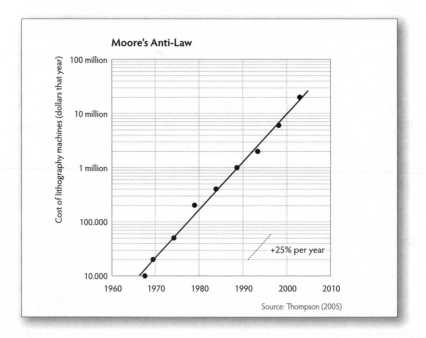

Moore's Anti-Law

Source: Thompson (2005)

As computer processors become more complex, their components steadily shrink in size. Each new generation is more difficult and considerably more expensive to manufacture. So far, that effect has been consistently offset by scaling up production. We may now be approaching the point, however, where that is no longer possible. Around the world, there are only a few production sites left that can still afford to invest in the next generation of microchips, which means they are likely to be even more expensive. *Source*: Thompson, S. E., and Parthasarathy, S. (2005). Moore's law: The future of Si microelectronics, *Materials Today*, 9(6), 20.

same time, the entire ecosystem of the semiconductor industry is changing. "Worldwide, only a small number of chipmakers can afford the most advanced processes. They are either microprocessor manufacturers such as Intel, IBM, and AMD or memory manufacturers such as Samsung. An interesting case is the emergence of superfabs such as TSMC functioning as 'silicon foundries' for many smaller manufacturers serving the consumer market such as NXP, ST, INFINEON, who become fabless or fablite. They focus on design of advanced chips and leave the manufacturing to the foundries. As for wafer steppers—the machines that make the chips—only one manufacturer, ASML, has kept faith with Moore's law."

De Man is convinced that the exponentially rising costs of each new generation of microprocessor chips generate diminishing returns. "A processor that's twice as powerful doesn't create twice as many possibilities. I've been using my PC for word processing and to create presentations for 20 years now. Processors have become a million times more complex over that period, but I still use my PC to do the same things. The word-processing software has picked up some additional features, but it certainly hasn't kept pace with the increase in computing power. That's the case with all electronics. You need an unbelievable amount of extra technology to achieve a small increase in usefulness. Further downscaling of transistors will only remain useful as long as it allows to reduce the cost per function."[4]

"A lot of companies have already stopped trying to make their processors any smaller," Hugo De Man confirms. "DVD-players or GPS devices don't need smaller processors. All that matters is the price, which goes on rising if you choose to stay in the race. Moore's law and making a profit are mutually exclusive. I wouldn't lose any sleep if Moore's law quietly faded away. Chips are small enough and powerful enough already."

NEW DEVICES

"Indeed, if I had a few billion euro to invest in electronics with the capacity to improve our lives, I'd certainly be looking elsewhere. I definitely wouldn't invest in further downscaling of transistors; the benefits are too limited. Other kinds of improvement are more important to many applications," Hugo De Man believes. "Hooking up electronics more effectively to their surroundings, for instance. You can incorporate sensors in a chip to detect movement or measure the temperature. Or little motors to move things around. That would give you devices that can monitor your heartbeat, say, or regulate chemical processes."

This is how information technology is rapidly evolving from the PC to the world of "ambient intelligence," where emphasis is more on smart communication and connectivity than on raw computing. Digital assistants in our pockets are starting to appear that provide at any time and any place a gateway to all people and information in the global village. They are increasingly able to connect and interact with smart objects in our surroundings and change the way we experience our environment. For these consumer-oriented applications, low cost and low energy consumption are more essential than supercomputing capacity. This leads to the development of new "more than Moore" technologies besides upscaling. These technologies involve, for example, a myriad of sensor techniques that are now being integrated in computer chips.

The science of electronics itself has also taken on new and different forms. You can now make electronics with such materials as polymers. That allows for flexible foils on which electronic circuitry can be printed, enabling you to literally roll out electronics in huge quantities. And we now have methods for packaging electronic components and sensors into cubes measuring less than a few cubic millimeters. "That's small enough for a device: The challenge now is to find smart ways of integrating sensors on a single chip," De Man says. "That would deliver spectacular advances in convenience, safety, and health. It would give us electronics that are truly aware of their ambient surroundings and can even control them."

De Man has, for example, proposed small devices that continuously monitor epilepsy patients. The idea is to analyze electric signals from the brain to detect the onset of a seizure and then take corrective action to prevent it from occurring. "That development is within reach right now; we don't have to wait for the next iteration of Moore's law. What we need instead is close cooperation with medical and psychology specialists—truly cross-disciplinary skills. Similar devices could be developed for other diseases, too." A related trend is the integration of electronics and chemical analysis along the lines of the "lab on a chip" described in chapter 2.5. "That will lead to devices that can analyze a drop of blood and take immediate action. Developments like this will promote preventive medicine rather than drug-based therapy. They have the potential to change the pharmaceutical business fundamentally, as they will reduce the consumption of drugs."

So we are seeing a new breed of devices that combine computing power with sensors. Rather than constantly seeking to make chips faster, more powerful, and more complex, new functionality could be added by making them more diverse. The resulting devices could then be used for applications that were previously resistant to computerization. The sensors and micromachinery of these new devices will create totally fresh ways of interacting with computers and, Hugo De Man believes, will revolutionize a number of fields. "It will profoundly change our industry, transportation, and the way we interact with our environment."

3.2

MORE COMMUNICATION

Things were very different 20 years ago. There was no Internet and no e-mail. The first text message had yet to be sent. Many European countries were still opening enormous transmission towers to put the finishing touches to their national television networks. Go back another 20 years, just as the first push-button phones were hitting the market, and a single computer would have taken up an entire living room should anyone have ever considered installing one. International phone calls were so expensive that people often timed them with stopwatches. The world has shrunk considerably since those days. E-mailing a research report or chatting online has become second nature. We can collaborate with someone on the other side of the world almost as easily as we can with a person two streets away. Companies use the Internet to outsource their accounts to India. Photographers sell their work all over the world. And if we want to, we can listen to Japanese radio in our European offices. Much of this book was written far away from the experts we interviewed. Yet in all the hundreds of phone calls, e-mails, and video sessions that went into its production, nobody paid the slightest thought to the physical distances separating us.

As the world shrinks, the way we use our communication networks intensifies. The volume of data we send is doubling every year, and the capacity of computer networks and telephone cables inexorably increases, too. Communication technology continues to improve at a rapid rate. And with each doubling of capacity, the price of transporting information halves. Things will no doubt look very different again 20 years from now. By that time, for instance, regions that currently lack Internet access will have been connected.

The first signs of these changes are already apparent. Africans are playing an important part in computer projects set up around the world by volunteers. They are involved, for instance, in developing

Linux—the open-source alternative to the Windows and Macintosh operating systems. Projects like this give programmers the chance to take part in global technological developments. Probably, business will one day begin to seek out more African specialists, too. Technology is likely to go on breaking down barriers in this way. That's the pattern as communication opportunities grow: The faster and the cheaper it is to communicate, the more people can do it, eventually drawing them into the global economy. Communication and development are utterly intertwined. Several new communication technologies have made the world a smaller place. Mobile phone providers are racing to offer us e-mail and imaging services, which in turn place a further load on the fiber-optic networks that facilitate all that communication. Other developments include videoconferencing, online gaming, Internet television, and remote medical imaging. These innovations are behind the remarkable growth we are witnessing in data transport. The traffic at major Internet exchange servers around the world has been growing exponentially for many years now. And there is no sign of a slowdown in this torrential flow of data across the globe.

There are, however, severe bottlenecks in our communication networks, which face overload at central nodes. These hubs now play such a crucial role that they have become the Achilles heel of the networks. A single malfunctioning hub can bring an entire continent's communication traffic to a standstill. In January 1990, for instance, a bug in a New York telephone exchange cascaded through the network, resulting in dead telephone lines across large swaths of the United States. For fully 9 hours, the network's nodes gripped one another in a stranglehold, taking each other down again the moment technicians brought a server back online. More than 1,000 flights had to be canceled or delayed, and long-distance rail couldn't operate properly either. Many businesses closed for the day and sent their employees home. Only after an old version of the software was installed did the network come back on-stream.[1] This was the most severe breakdown of a telecom network in history. In the years that followed, substantial amounts of money were spent in the United States and elsewhere on researching reliable telecom networks. This undoubtedly reduced the risk of similar cascades in future, but the crucial significance of network hubs means that constant vigilance is still required.

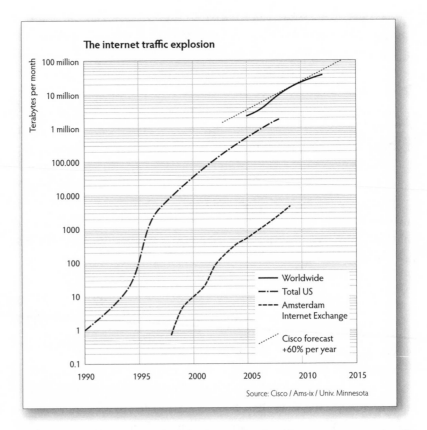

Internet traffic at key network nodes is growing by 60 percent a year, steadily intensifying the burden on the electronic servers that control it. Today's communication routing centers have huge footprints and a voracious appetite for energy. Breakthroughs in optical technology could help address this issue so that global communication can continue to expand.

BY THE SPEED OF LIGHT

Communication was revolutionized by the advent of fiber-optic cables in the 1980s. Charles Kuen Kao, who was awarded the Nobel Prize in 2009, launched in 1966 the idea that light could carry telecommunication signals via glass fibers. In that time however, it was not possible to manufacture glass fibers of more than a few meters long. The idea that light signals could be carried via glass cables was boosted by inventions in the 1960s and 1970s that enabled them to improve optical characteristics. Modern optical cables can now link two points up to 150 kilometers apart without requiring an amplifier. That's far more than copper electrical cables can manage. What's more, glass experts have come up with a clever trick for intensifying the light signal at the halfway point without having to break into the fiber. This makes long-haul fiber connections possible without the need for complex electronics en route. The same period saw the development of the laser, which provided a powerful light source for use in glass fibers. And the advent of compact and efficient semiconductor lasers gave fiber-optic technology an additional boost. As the years went by, it was discovered that more and more data could be squeezed down a single strand of glass fiber. The speed with which lasers fire their light pulses steadily increased, and the introduction of different-colored lasers meant that several kinds of light could travel down the same fiber at the same time. The big transatlantic connections currently use forty different colors, producing a combined capacity of over a terabit per second over a single fiber. In the lab, a single fiber is capable of transporting 26 terabits per second, enough to send the content of the entire Internet in less than a minute.

This impressive progress rapidly made copper cables, microwave links, and communication satellites old-fashioned. When glass-fiber links were first employed commercially in the early 1980s, they were used for the backbones of the networks, which carry the heaviest traffic. A global network of fiber-optic cables rapidly extended across the world's ocean beds. At the height of the Internet boom, several new cables were laid across the Atlantic and Pacific oceans. The system known as Africa ONE, meanwhile, has been encircling that continent since 2000. We are currently witnessing a new boom in optical cable laying that will lead to dozens of new undersea fiber-optic cables by 2015. This will create a truly global network with hundreds of thousands of kilometers of optical cable, each capable of transporting many terabits per second. Meanwhile, copper is also progressively giving way to glass fiber in the capillaries of the network. More than 1 million American homes are already linked directly to a fiber-optic network, and it won't be too long before we have gigabit links to our front doors—enough to transmit this entire book in a few milliseconds. Optical communication technology will therefore lie at the heart of the ongoing information revolution in the twenty-first century.

EVOLUTION OF THE INTERNET

The rapid growth in connection capacity meant that the structure of the networks has gradually changed, too. We can see this most clearly with the Internet, whose hubs originally played a much more minor role than they do today. The soldiers who conceived the technology that became the Internet were keen to avoid giving the network any kind of vulnerable central control. Instead, every computer would be able to decide for itself via which of its neighbors to transmit its information, passing it from one to the next, with each choosing the next step in the route. The data thus travel link by link to their destination, each step occurring so rapidly that there appears to be a continuous flow back and forth between two computers located some distance apart.

This decentralized structure meant the Internet was able to grow extremely rapidly. Anyone could connect very easily simply by hooking up a cable to the nearest Internet participant. Nobody's permission was required, and no adjustments had to be made elsewhere in the network. The same structure also made the Internet extremely robust. If a connection were to fail somewhere, the neighboring computers would decide autonomously how to divert the data traffic. In other words, the network had the ability to reorganize itself. The Internet was designed to be self-repairing, self-organizing, and self-learning. Such was the pace of growth, however, that certain computers began to assume a more important role than others. Perhaps they were located close to a transatlantic connection, for instance, or they happened to have some larger users nearby. It quickly became attractive for new connections to link to these more important computers as well. As a result, the network rapidly evolved toward its present hierarchical structure in which a limited number of large hubs occupy a crucial position.[2] In terms of communication speed, that is indeed the most efficient structure. We have now reached the point, however, where the hubs have become so important that they risk overload.

HUBS ARE HEATING UP

There are real problems today at the network hubs where data flows converge. Although copper has gradually given way to light in long-distance connections since the 1980s, electronic systems continue to hold sway within network hubs. To process the data at these hubs, light needs to be converted into an electrical signal. The bits and bytes are then examined, sorted, and turned back into light for transmission to their destination. Specialized computers known as *routers* run around the clock to keep the Internet traffic flowing. The importance of these machines is evident from the buildings in which the hubs are located. Security is paramount: That's obvious from the oversized plant holders in front of the main doors, which are solid enough to repel a truck. Visitors are routinely fingerprinted. Inside, the air-conditioning operates with a noticeable hum. All those routers get hot, and it takes a lot of power to keep the electronics below the boiling point. The cooling systems take up at least as much room as the routers themselves. Thousands of batteries stand by to deliver a few minutes of emergency power in case anything goes wrong. The routers in these buildings handle data at an incredible rate. The time needed to route each chunk of data has decreased by a factor of almost 50,000 over the past four decades. Today's network nodes have only a few nanoseconds to process each package of data (with a maximum of 1,500 bytes). As the capacity of communication links increases inexorably, the electronics at the nodes must speed up, too.

The capacity of communication links is developing faster than the microelectronics, which is a problem given the essential role the latter plays in our network nodes and hubs. The point will come when the electronics will no longer keep pace with fiber optics' annual doubling in performance. Electronic performance doubles only every 18 months (Moore's law; see chapter 3.1), which means electronic miniaturization and optimization can't keep pace with the growth of optical data networks. The gap between fiber capacity and electronic capability is therefore widening fast. If we reach the point where the electronics can no longer keep up, growth will cease. The situation is even worse at the largest hubs because these are the first to employ new broadband links, making them an even more attractive connection point, which further accelerates traffic growth at these locations. So what can we do to prevent our hubs from stalling? We asked Harm Dorren, director of the COBRA Research Institute and professor of optical signal processing at Eindhoven University of Technology in the Netherlands. Dorren is one of the scientists working to solve the hub problem. The optical switches he's developed in the past few years are among the world's most advanced. "The key problem is the enormous amount of energy that electronic routers consume," he says. "It's a real waste. It is caused by the enormous overhead merely to give the data a different direction. Routers have to decode each bit individually. It's like unloading a truck to read all the labels on its cargo and then loading everything back onto another truck. It's all hugely inefficient." The latest routers consume more than a megawatt of power. That's a lot, especially when you work out the amount of energy needed to process each individual bit. It takes 100,000 times as much power to switch a single bit as the bit itself actually contains. And that's just for the electronic processing and routing. As a rule of thumb, every processing watt requires another watt for the cooling system. "Talk about using a sledgehammer to crack a nut," Dorren says.

"Energy consumption is a fundamental problem with electronics," he continues. "If you want to speed up, the transistors have to move electrons around faster. Doubling your speed roughly means quadrupling your energy use. It's a simple law of physics. To limit power consumption, it helps to make the electronics smaller. That way, the electrons can be shifted over smaller distances, which is faster and consumes less energy. But since the size of a single transistor is approaching atomic dimensions, future downscaling of electronics becomes problematic. So increasing speed means using more energy. If you want to get rich quickly, you could do worse than sell fans and cooling elements for these systems." There are similar concerns with electronic signal processors, which is why the clock speed of the processors in our PCs has stopped rising. When faster than a few gigahertz, they become too hot to use (see chapter 3.1). Increased energy consumption for data routing causes several major problems. It not only means significantly higher power bills but also increased emissions from electricity generation and a further strain on the grid. The Internet already consumes more energy than global aviation.

TOWARD OPTICAL PROCESSORS

Around the globe, scientists are looking for solutions for the problems that have arisen at key Internet nodes. Approaches vary from creating different network structures to developing new computer protocols or altering hub components. Harm Dorren's strategy is to replace some of the hub's electronics with optical circuitry. He and his colleagues are working on components capable of routing light directly without relying on electronics. "You can, for instance, manipulate the color of a light pulse," Dorren explains. "In other words, you switch a pulse of light from one color channel to another. We were able to show error-free switching using such principles at bit rates approaching a terabit per second. The energy use of these optical switching principles is independent of the bit rate. In other words, if the bit rate is increased with a factor two, the energy per bit is reduced by a factor two."

"We focus on breakthroughs in optical circuitry that would make faster switching possible. Optical elements have the potential to do this without generating progressively more heat as with electronics. But it's a wrong idea to substitute electronic devices for their optical counterparts without changing the architecture of the systems." A key issue with respect to electronic architectures is that every bit needs to be processed separately. As a consequence, one needs to place millions of devices on a single chip. Replacing that by optical circuitry, Harm Dorren thinks, would quickly bring us face-to-face with important size issues. The reason is a fundamental one: The wavelength of the light used for fiber optics is 1.5 micrometers, which immediately determines the minimum possible scale. This phenomenon is referred to as the *diffraction limit*. Breaking this limit has its price. This was recently shown with devices that have dimensions that break the diffraction limit. These devices exhibit very high losses and often operate at temperatures far below room temperature. This makes application in real systems difficult. Another way to increase speed is to use different switching techniques, so that no longer every single bit needs to be inspected. This way, fewer processors are needed.

Mass production techniques will also need to be developed if optical circuitry is to succeed. Electronics likewise had to develop over a considerable time before the necessary techniques were mastered and production could be scaled up. The mass production of microelectronics, for instance, uses wafer scanners, which place all the electronic components on thin sheets of silicon. Superfast bonding machines then add the electrical connections. By contrast, connecting an optical chip to a fiber-optic cable still calls for extreme manual precision and skill; highly paid experts have to connect the fibres to the chip one by one. That may not be such a problem if you only want to make a few chips for use in laboratory testing, but mass production will require the development of new technology.

THE NEED FOR OPTICAL MEMORY

Several breakthroughs are necessary, therefore, before routers start to feature optical signal processors. And there is another big hurdle, too: A router needs not only a fast processor. If two packets arrive at the same time and have to be relayed via the same channel, one of them has to be held back temporarily. This suggests that storing data is of key importance. Existing routers with their electronic memories use the multigigabyte memory chips that are nowadays common in computers. They need billions of memory cells, but these can readily be fitted onto an electronic chip. "You simply can't do the same task optically," Harm Dorren points out. "Light has serious limitations when it comes to building memories." Dorren and his colleagues have tried a variety of techniques to come up with an optical memory. Loops can be used, for instance, to slow down a light signal in a fiber-optic cable. The loop forces the light to make a detour so that it arrives slightly later. But this would require a lot of fiber—millions of kilometers, in fact, for a fast router—making the idea a nonstarter. You can also try to transmit the light a little slower, as its speed depends on the properties of the medium through which it is traveling. Some materials can slow light by a factor of a million. That's at the expense, however, of the quantity of data that can be transmitted. And it still requires a light path measuring dozens of kilometers, which can't be integrated on a chip, and slowing down light costs a lot of energy.

Another way to make an optical memory is to design an optical switch. Scientists at Dorren's COBRA institute have done this using two small lasers acting in tandem as an optical "flip-flop"—a component with two stable states, which can therefore function as a memory cell. The flip-flop is a vital element in electronics. This latest optical version uses far less energy than previous attempts and offers extremely fast switching (measured in picoseconds). The flip-flop is so small that 100,000 can fit on 1 square centimeter—a serviceable number, even if it can't compete with microelectronic memory. But still you'll also have to reduce the amount of memory you need for your routers. You'd have to cut it by a factor of at least ten and probably more.

Recently, Dorren's group started investigating a different solution. Instead of buffering packets with optical means, one of the packets was converted to a different wavelength. Thus, instead of buffering data, a different color was used to mark different packets that arrive at the same time. "We computed the amount of wavelength converters that are required in an Internet hub node, and we think it can be done. We also learned recently that the problem of simultaneously arriving packets is not so much a buffering problem but a control problem. And both control issues and wavelength conversion issues can be solved today."

We'll probably never replace electronics with optics on a one-for-one basis. The fast speed and low power consumption of optical technologies are promising, but we'd need to address their weaknesses, too. It will probably never be possible to build processors as large as those we're accustomed to achieving with microelectronics. "This could mean that we have to organize the entire network differently," Harm Dorren concludes. "So the optical revolution would also require the fundamental rethinking of telecom networks from an architectural point of view." Solutions in one field of research risk causing bottlenecks in another. Yet conversely, problems in one area could be solved or sidestepped by switching to a different technology or research approach. Telecom networks are complex systems made up of elements that can strongly influence one another. You never know when packets of data are going to be sent, which means queues can build up at the hubs. "If you want to avoid collisions at the hubs, you need a lot more coordination. It's exactly the same with cars and expressways," Dorren says. "If there was some kind of supercontroller deciding when we could set off on a journey, it would solve our congestion problems at a stroke."[3]

The lack of centralized control was seen as an advantage when the Internet was being designed. Now, however, it has become a problem, as it makes it very difficult to change things. "Replacing the network cards in all the world's PCs clearly isn't an option. The most we can hope for is greater coordination on the Internet's busiest routes. But even that would be a major challenge." Dorren warns. "That's why interdisciplinary research is so important. You can't just change components; you have to think about networks and protocols at the same time."

HYBRID RECIPES

Meanwhile, microelectronics and optical technology continue to battle it out on several fronts at once. Electronics advocates believe the obstacles to faster, smaller, and more economical components can be overcome, whereas optical specialists like Harm Dorren continue to champion their solution. Both have good arguments, and both techniques are being developed in parallel as is so often the case with technology. In all likelihood, there won't be a single magic recipe that brings a breakthrough for the Internet's overheated hubs. Whatever solution ultimately emerges is likely to have a number of different ingredients. In practice, therefore, optical technology is not going to replace electronics across the board; each technology has its strengths and weaknesses. Electronics still wins when it comes to miniaturization, but optics excels in terms of speed and low energy consumption. For that reason, a fully optical telecom node is unlikely to replace its electronic counterpart except in situations where light is genuinely much more convenient. We're going to carry on using a lot of ordinary electronics at these network hubs. We'll probably keep a lot of electronic components for buffering and for the slower processes that support an optical chip. But the fastest components will probably become optical. We may be able to raise switching speeds to the terabit level that way. A first step toward a mixed solution is the development of hybrid electro-optical systems using optics for speed and electronics for storing information and calculations. A similar approach is also likely to be adopted in future generations of signal processors in computers.

In the meantime, when traffic at the hubs really begins to stall, the network will evolve toward a different topology. The air traffic network has been through a pattern of development similar to the Internet in recent decades. As air traffic grew ever more efficient, certain airports evolved into hubs at which large numbers of flights converged. Airlines seeking to establish a new route preferred to do so from a hub of this kind. The result was such severe congestion that centralized hubs became unattractive about 10 years ago, at which point it became more favorable to establish links between smaller regional airports. The flight network has since changed radically and taken on a much more anarchic structure reminiscent of the early days of the Internet. There is a similar trade-off in a communication network between the processing power of the hubs and the potential of the new connections. Hubs may be relieved by creating additional connections at other locations. That's just one example of an architectural evolution.

GLOBAL THINKING

Network capacity will continue to grow for the foreseeable future, helping to open up new opportunities just as we previously saw with the advent of telephony and e-mail. Several new applications already exist but are being held back by the "irritation factor" of networks that are too slow. You're not going to download a video if you have to wait an hour. Everything has to be really fast; otherwise, we won't use it. Network capacity is growing so rapidly that unlimited video communication will become possible, making life more enjoyable for grandparents in Paris, say, watching a live video of their grandchildren on vacation in Miami. Communication technology could significantly enlarge the world of elderly people (chapter 4.4). It will also be possible to attend scientific or business conferences via video link, watching presentations and asking questions as if you were there in person.

Yet communication technology has a much broader significance, too. To give just a few examples, it is a key to global education (chapter 5.1), a tool for bringing nations together (chapter 5.6), a way of making our cities more habitable (chapter 5.3), and a passport to enhanced medical care (chapter 4.1). Communication technology features throughout this book as a key instrument for improving the fate of humanity, as it renders geography increasingly irrelevant. There is much more to communication than simply doing business: It also affects the way we think, talk, and act. Communication brings us together. Nobody expects France and Germany to go to war with each other again; they have grown far too interconnected for such a thing to be conceivable.

3.3

REACHING EVERYONE

The majority of Earth's inhabitants don't have a telephone or e-mail, and many places still lie beyond the reach of established communication networks. That severely precludes development, meaning that new insights percolate through slowly and that essential services such as water provision are based on imprecise information. Lack of information exchange is also an obstacle to improving agriculture, education, and many other fields. Even in well-connected areas, networks are less dense than required to improve our safety and well-being. You need a fine-meshed network if you want to keep on top of the genesis of earthquakes, floods, climate change, and many other unstable systems. Sparsity in networks means a lack of control. Making network coverage more complete could help develop and stabilize our world. However, any attempt to extend networks raises significant problems, which we explore in this chapter using the example of radio networks. We should keep in mind, though, that similar phenomena are evident in other networks as well, including the power grid and social networks geared to education.

At first sight, radio is an excellent technology for filling the gaps left by other communication technologies. Its capacity is restricted, however, because wireless communication is limited by basic laws of physics, obliging broadcasting and communication companies, for instance, to battle it out for their slice of the ether. The number of radio and TV stations, mobile phones, and satellite connections increases inexorably, filling up every patch of the electromagnetic spectrum that can be used for radio. The historical pattern has been to develop new techniques and then to lay claim to unused spectrum. The old medium-wave bands were the first to be used for radio programs. Broadcasters later adopted FM, exploiting higher frequencies at the price of reduced range. New semiconductor electronics then became faster and brought access to even higher frequencies. Once the first global standard for mobile communication (GSM) slots had been filled, new space was opened up at double the frequency. The latest third-generation (3G) mobile phone technologies—like the universal mobile telecommunication system (UMTS)—work with higher frequencies still.

The higher the frequency, the faster the electronics you need, and the greater the capacity you get in return; that's the straightforward natural law of the electromagnetic spectrum. At frequencies around 60 gigahertz, you can provide every user with several gigabits per second—more than enough for two-way high-quality video. However, there is a price to pay for higher frequencies: They reduce your range. High frequencies are more easily absorbed by the atmosphere due primarily to the water in the air. Obstacles also absorb higher frequencies more easily. The usual strategy for dealing with this is to divide the area to be served into smaller cells, enabling you to install more transmitters without them interfering with each other. At the highest frequencies, radio waves gradually begin to behave like optical beams, including a growing inability to penetrate obstacles. This further limits the size of cells. For the highest capacity, we will ultimately have to reduce the range of each antenna to a single room.

There is clearly a trade-off in radio networks between capacity and range. As cells shrink, they serve a smaller area, but at the same time, they provide more capacity per user. In fact, some experts believe that 3G cell phone technologies still use frequencies that are too low so that cells are relatively large, and insufficient capacity is available in densely crowded areas, where it has to be shared among a lot of users. The relationship between capacity and range has changed as technology has progressed. Techniques are constantly being developed to exploit finite network resources more efficiently. The bandwidth needed to carry sound and video can be reduced so that it uses less frequency space without any reduction in range. New techniques for interactive digital television, for instance, use less spectrum than traditional analog TV broadcast methods, which is why analog broadcasts are being discontinued in many countries, freeing up space for additional digital channels. A wide variety of other techniques have also been developed in recent decades that make better use of the available bandwidth. Transmitters can hop to other frequencies when that's more efficient, and they spread their output across a greater number of frequency bands so that interference can be filtered out more effectively. The wireless (wi-fi) computer networks currently used in many homes and workplaces are also based on this approach.

THE INERTIA OF LEGACY

But there are much more radical ways of creating space. "If you turn the dial on your radio, you'll find whole stretches of silence," says Simon Haykin, who has spent a career researching smarter ways to use the broadcast spectrum, helping to improve radar systems significantly in the process. He is now professor at the Cognitive Systems Laboratory at McMaster University in Canada. Lots of space in the radio spectrum is empty, Haykin notes. Some are only used for part of the time, and others are used intensively. The spectrum has been allocated very precisely to all sorts of users, but they don't always utilize it efficiently. Huge holes in the spectrum remain as a result. "It's mostly silence," Haykin continues. "We could increase capacity significantly if we regulated it differently."

The problem lies in the centralized and rigid regulation of band usage. Governments have imposed strict rules to prevent transmitters from interfering with one another. Consequently, they have to be spaced well apart so that there is no interference even under the most unfavorable conditions. People might want to tune in from the depths of a building so transmitters must broadcast at high power to penetrate all that steel and concrete, further limiting capacity. The priority is to ensure interference-free reception in the most unlikely places, and so radio communication ends up being geared toward extreme situations that don't apply to the vast majority of users for the vast majority of the time. It all places a huge burden on capacity. A more intelligent approach would free up huge amounts of bandwidth and enable us to start using all those holes in the spectrum. There's no reason, for instance, why a different user shouldn't temporarily occupy a piece of spectrum while the actual licence holder doesn't need it. "The second user would have to be very flexible, though," Simon Haykin warns. "You'd have to be able to shift immediately to another place in the spectrum if the main user suddenly needed its bit again."

Although systems of that kind are currently being tested, replacing a substantial proportion of the world's existing stock of radio sets is a nonstarter. This legacy technology makes it hard to change the system. Classic transmitters and receivers have specialized hardware to decode the signals, which is different for each standard. A radio set needs specific components for frequency modulation (FM), for instance. If you want to convert to another technique like amplitude modulation (AM), you'll need to get out your soldering iron. Once a radio standard has been conceived and agreed, it is very difficult to modify, as lots of receivers would be rendered useless. To avoid replacement of all receivers, engineers often try to design a new radio technology that is compatible with the old one. By the time stereo FM transmissions began in the early 1960s, a great many FM receivers were already in use. Consequently, the stereo signal had to be cleverly concealed within the FM signal so that existing mono receivers wouldn't be adversely affected. This can be done but often at the cost of extra capacity or loss of quality. More flexible technologies can be employed in new, hitherto unused patches of radio spectrum. But you can be certain that these new technologies will also become outdated one day and that replacing them with even more efficient technologies will be no less difficult.

THE RECEIVER DECIDES

Simon Haykin has thought about this problem and sees a solution in a new type of radio in which dedicated physical components are substantially replaced by flexible devices that can be adapted via software—hence the name "software-defined radio." This would make radio technology much more flexible, as switching to a different standard would simply be a question of an instant software update. Ideally, this would occur automatically via the airwaves without users even being aware of it. It could hugely accelerate technical development and new applications. Transmission techniques would no longer need to be scrutinized during protracted negotiations on every update. Equipped with the right software, effectively designed radio chips could hop from one frequency to another and switch to new transmission technologies as they become available. They would also be more multipurpose and could therefore be produced in larger volumes.

Software can make radio receivers more intelligent, too, or "cognitive," as Haykin puts it. The equipment can be instructed to keep track of its surroundings in a manner reminiscent of human cognition. "The receiver rather than the transmitter should be central," Haykin says. "It can evaluate the features of its surroundings and detect interference and the strength of radio signals. The receiver can then communicate that information to the transmitter, allowing it to adapt accordingly. In other words, the receiver decides how the transmitter is going to transmit." Haykin has thought all this through in great detail, as a lot of things must be considered. What would happen, for instance, if someone deliberately tried to squeeze out other channels? Procedures would need to be established to stop the airwaves from being ruled by the strongest. The stability of the system is another issue: If radio transmitters all reacted to one another, the consequences might be unpredictable. A small shift in one location could potentially trigger an avalanche of frequency changes and output adjustments. A society full of cognitive radios would form a complex system with all that implies. "That kind of derailment can be avoided," Haykin thinks. "As long as all cognitive radios pursue their own interest, the system will remain stable. That's not based on my vision of society but on algorithmic computation derived from game theory."[1]

Cognitive radio could, in fact, be stabler than an intensively used monolithic network. Networks are prone to breaking down when radio communications come under stress—when there is a disaster, for instance, or simply a large crowd. A cognitive network could adjust to the new situation and permit only short-range communication, say, plus a limited number of long-distance connections. Regrouping within such a structure would be an efficient way of allowing information to percolate through quickly with minimum use of capacity. "Software-defined cognitive radio is a practical way to utilize the empty holes in the spectrum," Haykin argues. "It would make efficient use of the underutilized parts of the radio spectrum. The technology for doing so is available now. Mobile phone designers already use software-defined radio so that their products are capable of operating in lots of different radio systems; if a technique changes, the phone can be easily modified. Several armies upgrade their communication systems in the same way so that a technician doesn't have to go from one division to another to do it manually." The exploitation of radio spectrum doesn't only depend on technology, though. "The usable part of the spectrum has been precisely carved up as a result of complex diplomatic negotiations. No matter how cognitive or intelligent your radio equipment is, you can't just go out and lay claim to a bit of unused spectrum; even if you don't interfere with anyone, you would still be infringing international agreements. These will have to be amended if cognitive radio is to become an option."

Simon Haykin's ideas are a perfect illustration of how you can optimize complex networks by diminishing central control and allowing the individual elements to decide for themselves. That increases capacity and makes the network more robust. Haykin's research focuses on the realm of radio networks, but the philosophy behind cognitive radio is clearly applicable to other areas, too. Smart electricity networks, for example, hand over part of their central control to small decentralized units that make their own decisions. They learn from what they "see" in their immediate surroundings and constantly adjust their programmed behavior accordingly. Not only is that more

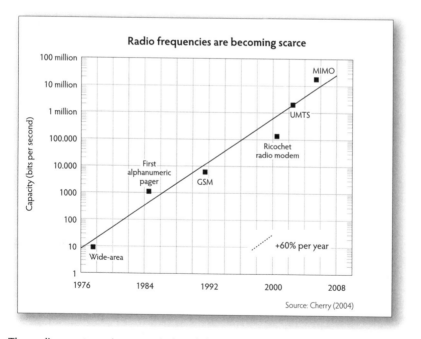

The radio spectrum is now strictly subdivided between users. It will be increasingly difficult to cover every corner of the planet unless we can identify new ways to share scarce spectrum resources. We are making far from efficient use of the radio bands currently allocated for communication purposes. We therefore need to find a less rigid method of radio frequency allocation. *Source*: Cherry, S. (2004). Edholm's law of bandwidth. *IEEE Spectrum, 41*(7), 58–60.

flexible, but it also makes the network less susceptible to outages. Air traffic control offers another useful example in this respect. It's worth looking at closely because it shows how well the cognitive concept can be applied in a totally different field.

FLYING LIKE A BIRD

Air traffic networks are geared toward extreme reliability at the expense of capacity. In densely populated regions, aerial congestion is the main cause of flight delays. As a passenger looking out your plane's window, however, you will be largely unaware of this heavy traffic because you rarely see another aircraft when you're up in the air. Passenger planes simply don't use the vast majority of our airspace; they fly behind one another on a fixed path. There would be plenty of room if pilots were permitted to leave those prescribed routes, but they aren't. Centralized management is already overstretched, and it couldn't handle the maze of traffic that would result if pilots were allowed to choose their own routes. The authorities also worry that more accidents would occur if aircraft were able to cross one another's flight paths. That fear is, however, groundless. After all, flocks of birds manage to fly in and out of each other's paths without colliding and without having to rely on central control. They simply are aware of their closest neighbors and make sure they take prompt evasive action whenever necessary.

It is interesting to see how the issue is being tackled in the case of aircraft. New flight systems have been developed that no longer require air traffic controllers at all. The idea is that pilots all monitor their own surrounding airspace and react to one another. In this new system, intelligent onboard electronics inform the pilot of any planes that are approaching in the opposite direction. If there is a danger of them coming too close, the system warns the pilot to move up, down, left, or right. The corresponding device in the approaching aircraft advises its pilot to change course in the opposite direction. The onboard systems of the two aircraft can't always contact one another, so the key element is to devise rules that would ensure that the aircraft always adjust their course in such a way as to avoid each other. This approach is much safer, as it shares the work of the air traffic controller among a large number of pilots, each of whom has only a small amount to do. It also makes travel a lot faster. Routes selected by pilots themselves are up to 10 percent shorter than the current network of fixed corridors and sharp turns.[2]

The new system isn't yet operational, but it will be much easier to introduce than it is to reform radio technology. The authorities have the power to compel the few thousand commercial aircraft that operate in the skies over a continent to adopt a different flight-control system. The numbers involved in the case of radio services, by contrast, are very different: You can't simply decree a fundamental change to an installed base of millions of receivers. Another advantage is that this approach allows a gradual evolution from central to decentralized network control. As a first stage, for instance, only aircraft with the new equipment would be permitted to deviate from the prescribed routes on the condition that they revert to the authority of the control tower as they approach an airport. Likewise, software-controlled receivers could be granted more freedom than others, allowing a gradual transition.

LINKING UNLIKE NETWORKS

The same approach can't be applied, however, when you're trying to get several different types of networks to converge and cooperate intelligently. This needs to be done locally, where the capillaries of the respective networks meet, so to speak. An example is the convergence of radio and fiber-optic networks, which are being linked more and more frequently. Routing decisions are required wherever this occurs. Fiber-optic systems offer much greater capacity, but they lack options when it comes to supporting mobility. The latter should instead be served primarily by radio, conserving scarce frequencies by transferring the signals to fiber at the nearest possible interface. A smart combination of radio and fiber optics will be the best solution in most cases. Simon Haykin's vision therefore needs to be broadened: An effectively designed, intelligent telecommunication system will benefit from the best features of each technology. Decisions in this area can't be made top-down. The specific characteristics of the different types of radio wave and optical connection will have to be considered at the relevant access point. The router will need to know that some frequencies won't carry beyond the room, whereas others can penetrate pretty much anywhere. It will have to take account of capacity, distance, and range.

Radio connections with the highest capacity and highest frequencies can be kept as short as possible by seeking the nearest access point to the fiber-optic network. Each room might, therefore, be equipped with a fiber-optic access point. For smaller capacity radio connections using lower frequencies, transmission can remain wireless over longer distances. If you want to communicate over very long distances, by contrast, the system will identify a fiber-optic backbone leading to your destination. As the signal approaches the receiver, it might switch back to a mobile link or continue via fiber-optic cable to a local computer. Using fiber optics to as great an extent as possible would significantly reduce the load on the radio-frequency spectrum. Connections like this already exist, but they tend to be inflexible, because they are often created for a single purpose. Incorporating intelligence would enable the network to respond immediately to changes, thereby saving precious bandwidth and making the system more robust.

Much the same is true where other networks join together. Integration of car traffic and public transport can only succeed, for instance, if drivers can organize their journeys based on actual transit times and maybe even have the capability of influencing timetables. The human network that brings education to underdeveloped rural areas, meanwhile, can only succeed if it interacts with labor and possibly even water supply networks. And finally, a logistic network is obliged to interact with local production networks.

3.4

CRYPTOGRAPHY

Electronic payments, Internet shopping, and mobile communication have fundamentally changed our society and not only because digital services have made our lives more convenient. Never before in our history has our behavior been tracked in such detail as it is today. The bank remembers precisely where and when we withdraw cash from the machine, the phone company keeps a list of all our calls, and the online bookstore knows exactly what we like to read. Stores use loyalty cards and discount points to record their customers' purchasing behavior. These databases have proved extremely useful. Companies can present us with attractive offers at just the right moment. All those data are useful for the authorities, too. They tell them if someone is in regular contact with a suspect or where a person was located at a particular time. The information helps the police and intelligence services prevent bombings or trace pedophiles. Much of this information is protected so that not just anyone is able to poke around our personal digital records. But protecting our privacy is increasingly difficult because the number of databases and communication channels continues to grow rapidly.

Messages have been protected since the beginning of written communication. For a couple of millennia now, military dispatches have been translated into an incomprehensible alphabet soup in case they should fall into enemy hands. Breaking codes was and is a challenge. The course of World War II would probably have been entirely different if military codes had been more robust. The process of encoding and decoding can, of course, be performed much faster and more effectively in the computer age than was ever possible with pen and paper. *Cryptography*—the science of encryption—has perfected its techniques over the past decades. Cryptographers are constantly searching for mathematical operations that will allow insiders to decode a message easily and make it so hard for outsiders that it's no longer worth the effort of even trying.

HOW CRYPTOGRAPHY WORKS

Several important security techniques use prime numbers—numbers that are only divisible by 1 and by themselves. Examples are 2, 3, 5, and 13 but also 7,901 and 16,769,023. All numbers can be expressed as the product of primes: 15, for instance, can be resolved into 3 times 5. Factoring large numbers into primes is a mathematical challenge. Without prior knowledge, it is unfeasible to factor very large numbers; only the most powerful computers in the world can factor numbers 100 to 200 digits long and only then with great difficulty. In 2010 the world record was the factoring of a 232-digit number, which took 2000 years of computing time, although it was actually done in a few months by a series of computers working in parallel.[1] Mathematical calculations like this lie at the heart of the computer programs that encrypt information. Outsiders who do not have the key can only decipher messages with extreme difficulty or not at all. An authorized user who knows the prime numbers used, by contrast, can complete the task in seconds. Software offering this kind of security is incorporated in browsers like Internet Explorer and Firefox, which we use to view Web sites. Related cryptographic techniques allow us to control our own data. Cryptography also allows data to be anonymized. The information in a medical file, for instance, can be separated from the patient's personal details. Privacy would then still be guaranteed if the file were passed from one department to another or if it fell into the wrong hands. Technology is there to protect our data, and it is used in a myriad of ways.

WHY CRYPTOGRAPHY WORKS SO POORLY

Our society is becoming increasingly dependent on cryptographic security. It not only protects our payment transactions but also our new passports and the music we buy from the iTunes Music Store. How safe is that protection? What would happen if a smart mathematician were to crack the codes? By no means every security measure

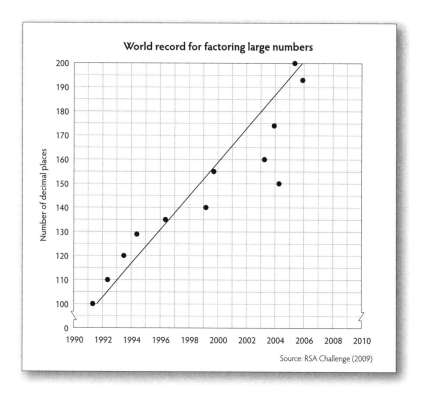

World record for factoring large numbers

Number of decimal places

Source: RSA Challenge (2009)

The protection afforded by cryptographic systems is based on the impossibility of factoring very large numbers. But security is a product of its time and needs to be constantly revisited. The challenge is to do this against a backdrop of steadily growing databases and increasingly complex data exchange. *Source*: The RSA Laboratories, http://vermeer.net/rsa

has proved impregnable. Indeed, cracking digital security has become an attractive sport for some people. Pay-TV chips, phone cards, and electronic tickets have all been hacked. The first DVDs had barely appeared on the market before the security method was published on the Internet. This was rapidly followed by software with which anyone could copy a DVD despite its cryptographic protection. France Telecom caught a number of fraudsters who, rather stupidly, made calls that cost more units than even the highest denomination of phone card could possibly cover. But a lot more fraud probably goes undetected because the hackers in question are a little smarter.

What problems with cryptography result in protection mechanisms that are so often fooled and broken? Henk van Tilborg, professor of cryptography at Eindhoven University of Technology, has studied security flaws in depth. Our conversation with him was the latest in a long line. We regularly discuss data-security issues with him, and each year, his tone grows a little more pessimistic. "Data protection has deteriorated in recent years despite the powerful options offered by technology," Van Tilborg says. "We already have the technology to virtually guarantee security. We've had cryptographic mechanisms of proven strength for many years now. The techniques exist; we just don't use them or we don't use them well enough."

Henk van Tilborg is one of the rare people who care about the personal data we give away to others. Googling his name produces only a long list of scientific papers and a LinkedIn profile. You might also turn up his lectures on keeping secrets at a children's university. But that's about it. He's not to be confused, incidentally, with the Catholic missionary of the same name. Most people, he notes, are considerably less cautious. "Nobody really cares. People don't seem to worry whether their neighbor can peek into their PC. Most people don't attach any importance whatsoever to effective data protection. We cheerfully hand over every detail of our shopping habits in return for a few air miles. You can even break into the systems controlling certain industrial installations without too much effort. That's unlikely to change until more people lose patience with all the errors and outdated information following them around or when some terrorist figures it out and then takes over a chemical plant."

Van Tilborg thinks that since ordinary people don't care about security, designers are less likely to focus on it either. "Many devices could be protected much more effectively," he confirms, adding that even when he and his colleagues have helped develop rock-solid security techniques, things frequently go wrong further down the line. "Cryptographic protection is often the final element of the design. It has to be cheap. That's why the chip in a phone card has little processing power and hence only limited protection. Designers prefer to use their budget to more appealing ends. They'd rather exploit a bit of spare printer memory to squeeze in another font or two. Anyone looking to use a small amount of processing power to improve security usually ends up on the losing side. Designers tend to sacrifice security in favor of added functionality. It's a question of choice. Really good security often requires a little extra processing power." The same syndrome is familiar from everyday life. New homes are often fitted with standard cylinder locks because better ones are expensive. You can get doorplates to protect the cylinders from being punched out and locks with built-in steel pins that aren't so easy to saw through. But people only tend to buy higher-quality locks after they've been burglarized.

Manufacturers save on security for rational reasons, Van Tilborg thinks. "Security doesn't sell. Protection is invisible, especially when it's doing its job. It doesn't make for sexy advertisements. Far from it: To highlight your product's excellent security features is to admit that there are dangers out there." Security features are often a hindrance, too; there are passwords to remember or special procedures to go through when you first hook up a new device. It is invariably at the expense of the ease with which the user can gain access. Adding security features costs time, too. In a world where every manufacturer is rushing to be the first to market, security is often only an afterthought. Leaving it out shortens development time. And even when security is taken seriously, it's often too late to change the course of development, and it ends up being added in a hurry during the final design stage. "Protection for DVDs was a rush job, as the film industry was divided over the new medium until the very last minute," Van Tilborg says. "But protection needs to be considered at every stage of the design. You have to invest the necessary time." Companies often get away with poorly secured connections because it is primarily others who have to bear the consequence of any data leakages. "In practice, you only see strong security teams at companies that directly depend on the protection of their products, such as Pay-TV channels and banks," Van Tilborg points out. "Elsewhere, designers interested in better security often have to battle within their own companies."[2]

GROWING COMPLEXITY

Security issues will only become more pressing in the years ahead, Henk van Tilborg thinks. More and more small devices need to be protected. Wireless communication between intelligent equipment is still in its infancy. Electronic organizers (PDAs) and mobile phones will increasingly communicate with computers, which in turn control wireless printers. Thermostats, refrigerators, and televisions will soon be communicating with each other and with your PC. Other information systems, too, are increasingly connected. Airlines, for instance, link their booking systems to car rental firms and hotels. Before long, the power company and the letter carrier will automatically know who is on vacation. All this communication will need to be secure because intruders will always enter through the weakest point. You don't want your neighbors to access your computer via your wireless printer or to switch off your freezer when you're on vacation. It's far from easy to ensure the security of this type of connection. Coupling increases complexity and therefore vulnerability. Every element in the system must be scrutinized. Complexity is the worst enemy of security.

According to Van Tilborg, security has become increasingly difficult as computers have grown more powerful. "That makes it easier for hackers to get in." The countermeasure is to use bigger prime numbers. But there are limits there as well. Choosing a number that's twice as big doesn't make security twice as good. "As computers get faster, you have to use disproportionately large numbers to rule out the risk of hacking. There are limits to that approach, especially for small devices. So you have to look to other cryptographic techniques, which makes the whole thing more complex." Applications require ever more complex cryptography, and there is no sign that this will stop in the decade ahead. The result is more work than cryptographers can handle. Despite the Internet boom and the proliferation of electronic data files, the number of cryptographers working in industry has barely risen over the past 10 years. As a result of growing complexity, security will steadily deteriorate, and hackers and fraudsters will experience less and less resistance.

WHO WILL PROTECT US?

The way to address the issue of growing complexity is through increased standardization, Henk van Tilborg thinks. "Every institution comes up with its own security system at the moment. We're overloaded with credit cards, debit cards, and other plastic. Tax collectors send out their own authorization codes, and banks issue us with special devices. Reducing that variety would make the cryptographer's task less daunting. The fewer systems they have to take into account, the better they can perform their tasks." He cites the example of the "digital companion" his team is working on. The device Van Tilborg has in mind resembles a phone. It communicates with the bank, the hospital, the tax department, and other institutions that work with sensitive data. "The information is kept in the device, so you retain control over it. The different organizations can then use the system to communicate. The tax authorities obviously wouldn't be able to look at your medical details: Each body would have its own key and specific authorization. Each one could also decide its own rules, including the right to amend the data in the companion. But you would only have to design and secure the system once." The challenge will be to convince all the different organizations to use the system. "The technology is there, but it will take a lot to get everyone onboard. It's tempting for each party just to go ahead and make its own system."

A central system like this would obviously be attractive to intruders. If you manage to hack it, then you'll have broken into everything at once. "But you could put all your efforts into protecting that one system," Van Tilborg counters. "At the end of the day, that's better than having a whole range of systems, each of which has its own specific weaknesses. Another important factor is that it would be a lot easier and more convenient to use than lots of separate systems. You only have to take one thing with you. It would make things very simple, and that's important. Otherwise, no one will want better security."

Is Van Tilborg's solution 100 percent secure? "There are several security methods we've been able to rely on for decades. They're based on solid mathematical premises. But you can never rule out the possibility that a new way will be discovered to crack them. We'd love to be able to prove that a particular technique is entirely secure, but that's very difficult mathematically. There's plenty of secure technology out there, and it doesn't happen too often—fortunately—that supposedly unbreakable mathematical techniques turn out to be unsafe. But you can't guarantee security forever. Research has to continue to make sure we're not caught unawares by new ways of circumventing security measures."

One dangerous development could be the advent of the quantum computer.[3] Although still in its infancy, its processors would work in an entirely different way, taking advantage of phenomena at atomic scale. This new breed of computers could perform certain calculations much faster than classic machines. "That would mean the end of a widely used security technique—the RSA algorithm," Van Tilborg admits. "But that's still a long way to go. Quantum computers can only count up to 15 at this stage. We have time to prepare for their development. People are already working on security systems capable of standing up to quantum computers." A more urgent focus is on the short term. "When it comes to a lot of practical security, huge improvements could quickly be achieved using techniques that are available already. It's simply a question of attitude and awareness."

WHY BOTHER?

Security is becoming more complex, the number of experts isn't increasing, and consumers don't think it's important. The technology is there, but it isn't being used. Henk van Tilborg's message certainly offers food for thought. We're all acutely aware of reports of intercepted credit card details, phished passwords, and plundered bank accounts. What if more confidential information were to leak? You needn't have guilty secrets to be concerned about that possibility. Even model citizens are aware that more and more is known about them. The hacking of computer data has implications for our own security. The easier it is to come by other people's personal details, the easier it is to pass yourself off as someone else. It used to be necessary to steal someone's ID card, assuming your country had such things; now you simply need someone's password or date of birth. Protecting your property means securing your personal data. There's a good reason we have locks on our front doors and we lock our parked cars.

Even if the information hasn't been hacked, however, we have less and less control over who knows what about us. Using the kind of coupled databases Van Tilborg has in mind, our own data will disappear from view. The more information is combined, the greater the likelihood of errors occurring. An incorrect note in the records of a credit card company can haunt a person for years, making it difficult to get a loan. Blacklists are frequently shared with other institutions, making it hard to correct a mistake or even to work out who has what information. And even if no errors have been made, you still have the problem that the information isn't complete. You only have to Google your own name to see what kind of distorted picture the outside world has of you. But it's hard to do anything about that either. A fair approach would insist on everyone retaining control over their own data. There's nothing new about this. There have always been people who have been misunderstood or slandered. What is new is that the dissemination of such misinformation is now so much wider.

Data will be preserved until far beyond our deaths. That means we're going to be confronted more and more often by our own history, which will impact our opportunities in society. Recruiters are already looking for all the available information about job applicants. The more they dig up out of old records, the less equal people's opportunities will be. Individuals who ever make a mistake will have to answer for it for the rest of their lives. Confidentiality and equality are intimately related.

Old records make people vulnerable if social attitudes change. History teaches us that democracy can be severely tested from time to time. And it is precisely at moments of heightened tension that we rely on the anonymity of the ballot box and must be most certain that we are not being manipulated or that essential details of our identity can be traced. In occupied Europe in World War II, it was considered an act of resistance to destroy the registrar's records, as that information made it far easier to identify and murder Jewish citizens. Democracy demands privacy, especially at moments of tension. We don't know how society will respond to the steady growth in the flow of data. Never before has so much been known about so many people. And it is more and more common for that information to be available for anyone to read without any form of protection. Lots of people cheerfully publish every detail of what they're up to on the Internet. Things they'd never tell their neighbors are there for anyone to read in cyberspace. Services like Facebook, LinkedIn, and Twitter have hundreds of millions of users, all of whom eagerly communicate about where they are and what they're thinking.

Looking back over human history, we find a steady increase in scale. News was mostly shared in agrarian communities within small groups. Growing urbanization was accompanied by an increase in scale and the number of cross-connections. In our current global village, we share everything with everyone. The scale couldn't be any bigger. What are the implications of all this openness about our everyday lives? What would happen if everyone were to take all that information seriously? Will social dynamism be increased by these waves of information about other people? Or will it become more rigid, seeing as it will be so much easier to identify like-minded people? What is the future of democracy if the government knows everything about us? Could you love a person about whom you know everything before you've even met? And what impact will this openness have on our identity? Will we hide our personalities away and reveal only what we want to? Or will we learn to forgive people's mistakes and understand that a person's past doesn't tell us anything about his or her future?

We'll learn the answers to these questions in the decade ahead. Scientists are currently researching online social networks to identify patterns and developments.[4] Studies like this could provide a clearer understanding of how our behavior will change under the influence of the mass. Our conception of privacy—always variable—is likely to change again. Greater understanding of the side effects can help us decide which information is to be safeguarded to protect our lives and to stabilize society. That will also help us make the most effective possible use of the scarce cryptographic resources being developed by Henk van Tilborg and his colleagues.

3.5

MANAGING FAILURES

Computers are the engines that drive our society. We get paid via computer, and we use them to vote in elections; computers decide whether to deploy the airbags in our car; and doctors use them to help identify a patient's injuries. Computers are embedded in all sorts of processes nowadays, and that can make us vulnerable. Because of a single computer glitch, large payment systems can grind to a halt. When computers malfunction, we risk losing our power supply, our railway links, and our communications. Worst of all, we habitually shift responsibility to computers and blindly follow their advice. This is why patients occasionally receive ridiculously high doses of a powerful drug or a car driver who blindly follows his satnav may end up in a ditch. Ubiquitous computer use can cause otherwise responsible people to leave their common sense at home.

We're all too familiar with poorly designed software, computer errors, or—worse still—programs that flatly refuse to function properly no matter what we do. It is hardly surprising then that computer failures cost the world hundreds of billions of dollars a year. In the United States alone, failed computer projects are believed to waste $55 billion annually.[1] And the media only report the tip of the iceberg— the foul-ups that cost millions or result in fatalities. For instance, in the 1980s, several cancer patients were killed by a programming error that caused the Therac 25 radiotherapy unit to deliver excessive doses of radiation. In 1996, Europe's first Ariane 5 rocket had to be blown up a mere 37 seconds after launch in what might be the costliest software failure in history. In 2007, six F-22 aircraft experienced multiple computer crashes as they crossed the date line, disabling all navigation and communication systems. The list can be extended endlessly, and there are many more failures that we never hear about. Only about a third of all computer projects can be described as successful, and even these are hardly error-free.[2] Why can't we prevent programming mistakes?

Could we improve computers and their software to protect society from the "moods"' of its digital machines?

SYSTEMATIC DESIGN

In many cases, pressure to cut costs and meet delivery deadlines is to blame. All the same, there is plenty of room for fundamental improvement. We would be much better off, for instance, if software makers adopted the kind of robust design methodologies we find in other technological fields. When architects draw up plans for a house, for instance, they begin by calculating how strong the foundations need to be, as it's not easy to change these things further down the line. Hence, exhaustive calculations are done before any kind of construction work begins. You get only one chance to do it right. When the moment finally comes to start constructing the house, it's simply a matter of executing the instructions and then testing to see if all has gone according to plan. The bricklayers don't have to figure out for themselves what color mortar the architect had in mind. The concrete people don't have to decide how much reinforcement is needed. That's the designers' responsibility.

Surprisingly, software is rarely designed in such a rigorous manner. Since computer code can be altered at any time, new ideas are frequently incorporated in projects already under way. Objectives and functionality are changed in line with advances made during the execution of the project. Working more systematically would be a step in the right direction: Creating complexity requires strict design procedures. Chip producers learned to adopt systematic procedures the hard way after it was found that Pentium processors were making calculation errors. The fault cost Intel in the region of $1 billion. The company's principal competitor, AMD, also felt obliged to introduce design procedures that were systematic and fully transparent. AMD now claims it can demonstrate that a design is correct even before it starts manufacturing.

Systematic software design is still the exception, and even then, it's only a first step. Software is far more complex than a house, so errors are inevitable, even with the most painstaking design methods. Identifying errors within the complex edifice represented by computer software is notoriously difficult. The applications used by a bank, say, can quickly run to tens of millions of lines of code, not to mention the fact that these lines aren't executed in a linear fashion. Computer software is a jumble of cross-connections and jumps to other parts of the program. It comprises a matrix of instructions with countless links. The structure of a piece of software is hundreds of times more complex than that of a skyscraper. It's advisable, therefore, to test the program once it has been constructed. Systematic testing procedures have been developed that replicate user input and check the functioning of the program under a variety of conditions. Similarly, newly constructed bridges are stress tested in some countries by loading them up with a line of trucks so that engineers can verify that the strain on the bridge's suspension conforms to the original calculations. This won't predict how the structure would stand up to an exceptional natural disaster, but it can help identify any miscalculations. Similar testing is rarely performed on software, however.

Since the early 1990s computer scientists have developed formal analysis tools that run through every possible cross-connection within the code. This kind of analysis can only be performed on a very limited part of a computer program—when testing a new computer communication standard like universal serial bus (USB), for instance.[3] But it does scrutinize the code from top to bottom and lists all the errors it detects. This makes the technique a genuine improvement. Similar methods may be used in the future for critical sections of code in software designed to control nuclear installations, military hardware, or financial systems. Yet they can never be applied to a computer program as a whole. Working through every possible permutation of computer instructions one step at a time isn't an option. There are far too many of them—more combinations, indeed, than there are atoms in the universe. It would literally take an eternity to work your way through all the cross-connections.

Thus, although we can do something to prevent software errors, computers will never be perfect. To banish errors more efficiently, we may need a radically different approach inspired by complexity science.

REDUCING COMPLEXITY

Klaus Mainzer is professor of philosophy at the Institute of Interdisciplinary Computer Science at the University of Augsburg and, more recently, the Munich University of Technology. We met him in the medieval German city of Augsburg to discuss new approaches to computer programming. Augsburg was a free city for many centuries, traditionally offering a haven to refugees from the centralized, monoreligious principalities that surrounded it. The old church and synagogue, for instance, are built alongside one another and even share a common wall.

Inside the modern university building, Mainzer is happy to talk about complexity. He originally studied complexity in the context of digital systems but has widened his study to the difficulties confronting designers and a broad range of nonlinear processes at work within our society. He is the author of a best-selling book on complexity.[4] Digital systems are becoming more and more complicated, he says. "Take the controls in your car: The amount of electronics is unbelievable, and it's growing all the time. The more complex they become, the more common it is for these systems to suddenly stop working for some unknown reason. Electronic systems and the software that controls them are rigid, inflexible, and strongly interrelated. The slightest anomaly can cause the whole car to fail. Every now and again, you pull up at the lights and everything stops working. You have to restart the car to get them to function properly again."

Problems like this are not unusual in complex systems. Local errors have also triggered widespread failure in telecommunication and power supply networks. "Rebooting" your car may help, but the real solution is to decentralize the controls, Mainzer continues. He describes a prototype car he built together with designers at Mercedes in which the vehicle was broken down into autonomous parts. "Each 'carlet' can configure itself and cooperate with the others. If part of the unit controlling the lights should fail, its other carlets take over what is, after all, a vital function. That's exactly how your brain works. If you suffer a stroke, certain functions fail. You might not be able to speak anymore. But you compensate by writing and using gestures. And it's often possible to reactivate the speech centers—albeit after a considerable amount of training. The system self-organizes into a new and stable configuration." Carlets have the added benefit that they are limited in scope and easier to test. "The same concept is also helpful when adapting software," Mainzer says. "You can easily extend the functionality of the car by adding new carlets. It's a totally different approach to software design. You don't need to build a single huge computer program that can't be readily adjusted. The system can grow in an evolutionary way instead, constantly regrouping whenever new functionalities are added. Evolutionary software architecture of this kind also makes it easy to change functionalities during the lifetime of a car."

A similar approach could make the microelectronic circuitry of computer processors more reliable. We're packing more and more components onto a chip, which makes it more difficult to implement them without any errors. If there is a single fault among all those millions of components, the entire chip will malfunction. The lives of chipmakers are increasingly defined by components that break down prematurely or do not function stably or reliably over time. Chip designers already spend about a third of their time trying to head off component failure. And these problems will only increase as miniaturization continues: The smaller the details become, the tougher it is to avoid faults. If a chip has a defect, it should be able to reprogram itself so that the fault can be bypassed. In the same way, a chip can take advantage of components that happen to operate particularly well by giving them a bigger role. In this way, performance is no longer determined by the weakest element but by the strongest. This in turn opens the door to further improvements. If chips have a flexible and intelligent internal network, they can be made to adapt their function when circumstances change. A similar approach also helps in telecommunication and power supply systems. A well-designed network should be able to bypass a fault and reroute the data stream or electrical power.

NEURAL NETWORKS

Klaus Mainzer has also studied another approach to computing: the use of neural networks. These crudely imitate the functioning of neurons in human brains, based on the pioneering work of American scientist John Hopfield in the early 1980s.[5] Hopfield linked up small processors that communicate with their neighbors in roughly the same way that brain cells do. One side of the network can be connected to an input device, such as a camera, and the network then communicates its decisions on the other side. The neural network is trained using hundreds of situations from the past. In this way, the individual components of the network learn how they are supposed to respond to the signals from their neighbors, and a correlation eventually arises between input and output. You don't have to write a program for a neuronal computer of this kind; instead, you spoon-feed it with examples of "correct behavior."

Neural networks can be deployed in all sorts of situations where it is hard to specify rules, Mainzer explains. "Neural networks can make decisions based on vague criteria in areas like controlling the temperature of a building or speech recognition. Learning and self-organization are the distinguishing features of a neural network. Like the human brain, neural networks are flexible and fault tolerant. That's why they have become such an influential paradigm in complexity research." Neural networks don't execute computer instructions one by one as "traditional" computers do; their processors work in parallel. "Or rather, we could build them to do so. For practical reasons, neural networks are still largely executed on standard computers in which they are simulated using a software-controlled system. The goal, however, is to build special hardware, as that would result in systems that are truly robust and adaptive."

"Today's neural networks are still far too primitive to even begin to emulate our human consciousness," Klaus Mainzer admits. "They only resemble our brains to a limited extent." A few thousand processors are active in the computer's network compared to the 100 billion neurons in the human brain. "They don't function in anything like the same way as living nervous systems. And they don't need to either. No significant progress has ever been made in the history of technology by slavishly imitating nature. Humans didn't learn to fly by copying birds; you can't just put on a feathered suit and take to the skies. Only when we'd figured out the laws of aerodynamics did it become possible to build a flying machine," Mainzer concludes. "If we want to build really intelligent computers, we're going to have to discover the basic laws underlying the functioning of the brain."

3.6

ROBUST LOGISTICS

Our lives seem to revolve around schedules. If we don't honor them with second-to-second precision, we miss our trains and our workplace rosters fall apart. We're reliant on one another, and we constantly have to coordinate our schedules with those of others. Planning is crucial to our industry, too. If you unexpectedly run out of nuts and bolts, you can't make any more cars, and the entire production process grinds to a halt. No manufacturer can afford that, so industrial companies employ large teams of specialists whose job is to ensure there are never any shortages of key parts. A worldwide logistic network has become our industry's lifeblood.

The central issue facing logistics is that of reliability. How do you keep your supply network intact? And how do you limit the consequences if it fails? These are questions that go far beyond the supply of nuts and bolts for new cars. Reliable logistics touches equally on the web of interactions that determine food production and the optimization of the Internet. It also extends to power supply, telecommunications, and workforce. Reliable networks make our society tick. But they face uncertainties of various kinds. That lends a broader significance to insights gained from industrial logistics, which offer us tools we can use to optimize networks and account for uncertainties in other areas as well.

INVENTORIES

The reliability of a supply network is intimately bound up with the inventories you need to maintain. Businesses hold millions of dollars' worth of supplies in their warehouses to make absolutely certain they never cease production due to a failure in the supply chain. So the key question is how large a stock do you need to hold of each component?

Smart planning to hold down inventory levels in your warehouse generates immediate savings. On the other hand, you need enough stock to ensure continuity should anything go wrong.

Optimizing storage is a common problem in supply networks. There is always a trade-off between the reliability of the network and the need for it to be profitable in an economic sense. Where local storage is impossible or very expensive—in a power grid, for instance—the network must be extremely reliable. Where local production is reliable and storage is cheap, you can take the pressure off a network by incorporating a local cache. Nuts and bolts are cheap, so industry is likely to maintain huge stocks of them, sharply reducing a plant's reliance on any given supplier. More valuable components such as engines, by contrast, have to be delivered "just in time," which requires a tightly managed procedure.

This is not simply a matter of transportation links between companies. Complex networks also exist within individual businesses for manufacturing and storing the intermediate products that provide a buffer for contingencies. Accordingly, effective management of logistic networks has to take into account many aspects of the relevant business. Personnel rosters have to be drawn up, machinery needs to be deployed as efficiently as possible, and customers' orders have to be fulfilled on time. Planning is an immense jigsaw in which everything depends on everything else. To do it without error takes powerful computers and different kinds of software, such as Enterprise Resource Planning (ERP) and Advanced Planning and Scheduling (APS) systems. These are often sizable and expensive systems linked to bar-code scanners, warehouse robots, and production machines. They precisely track supply levels, the location of each delivery, the status of orders, and machine availability within the plant. Together, this provides overall control of everything that goes on. Fast data connections enable a firm's computers to communicate with its suppliers, customers, and subsidiaries. Large corporations frequently combine data from all over the world to help them make the most effective operational choices.

EVER MORE DATA

Supermarket chains use this kind of advanced planning to stock their stores. Large computers in local branches collect checkout data, which are then processed at the head office to calculate precisely how many packets of coffee, bottles of cola, and bunches of bananas each branch needs. Scheduling the subsequent deliveries based on those data is, however, horribly complex. There is so much information, and the calculations are so computer-time intensive that even the most powerful computers can't handle the process perfectly. Planning the journey of a single sales representative is hard enough, as illustrated by the celebrated "traveling salesman problem." This sets the challenge of finding the shortest route between a number of cities. If there are fifteen cities, for example, there will be billions of possible routes. Mathematicians have devised strategies to tackle the problem, but it is very difficult to demonstrate that a particular route is indeed the best. Meanwhile, industrial planners face far more complex versions of this mathematical conundrum every day. Each situation requires a different strategy to arrive at the optimum solution.

What happens in practice is that the problem is simplified and assumptions are made, allowing the planner to use a standard software package that generates a usable schedule after a few hours of processing. The results are often far from perfect, but they're as good as computers can manage. There is steady improvement, of course, as computers become more powerful with each new generation of processors. So what will companies' operational planning look like 20 years from now? Will an omniscient computer plan every tiny detail? Will supermarket shelves be kept stocked to the last gram?

One scenario is that the software will be constantly refined to take better and more accurate account of all the available data. Companies will then perform their calculations with increasing precision, adding more detailed information and solving ever more complex variants of the traveling salesman problem. Data will become progressively more detailed as a growing amount of information is recorded concerning every aspect of a company's activity. Operations will increasingly be performed digitally, making them easier to monitor. Every phone call, computer process, and movement will be recorded and then used to improve planning so that personnel, machines, and suppliers can be deployed even more intelligently. More data and greater computing power equate in this scenario with more sophisticated planning. Ideally, every time a packet of coffee is sold, this will be communicated directly to the distribution center, which will take immediate account of it when preparing orders.

THE FUTURE OF PLANNING

We talked about this prospect to Ton de Kok, who is director of the European Supply Chain Forum—a network of global companies focusing on supply chain management. As professor of the quantitative analysis of logistics systems at Eindhoven University, De Kok is concerned with collecting and using large volumes of data. "Knowing everything about a business doesn't produce effective plans," he says. "More data and better software don't necessarily guarantee greater precision. Larger calculations merely give the illusion of precision. It's a totally wrong approach. Complexity theory suggests that it's simply not practical to find an optimum answer to most planning issues. You can go on doubling your processing power, but you'll still have to simplify; otherwise, it won't work. Planning computers use lots of rules of thumb, and that's not going to change, especially as the amount of data goes on increasing. The price you pay is that you abstract the issues to such an extent that the results you get no longer match the reality."

What's more, De Kok continues, the data you collect are often less precise than management thinks. "It's tempting to believe that more details will result in more precise planning. Businesspeople and scientists tend to believe that everything works in a predictable way. I don't agree. The future is never precisely what you expect, and that goes for business activities, too. Business operations aren't a chess game in which you can figure out every move right through to the end." This becomes obvious as soon as you start to measure production times with a stopwatch—something De Kok always does with his students, for whom it is invariably a revelation. "If you actually stand next to the person doing the work, you realize that things don't all occur in a predictable way; you get big variations instead. It makes a difference whether it's a Monday or a Friday. Some employees are more energetic than others. Some machines work better. Just try to predict how long it will take a truck to do a 100-kilometer journey between two cities. If you can get it right to within 15 minutes, you're doing well. Tomorrow they might dig up a road somewhere, and your delivery times go out the window. Nor is it possible to forecast demand. We simply do not know which customer will buy our products next week. The pitfall is to assume that your estimate of this demand is fully correct. You then base all your actions on a false assumption."

The problem, according to De Kok, is that planning software doesn't take account of this kind of variation and uncertainty. "Computers simply churn out ever more detailed plans. In many cases, the software will generate a schedule in which every last bit of slack—along with any potential flexibility—has been removed. In this kind of planning, no machine is ever left idle, and there's not a single bolt too many in the warehouse. Employees have to keep up a demanding pace to stay on schedule. The lack of flexibility means, however, that even a minor delay can have major consequences. That's the dilemma: The more precise your planning, the harder it is to adjust and the more frequently you'll have to throw it out altogether. The implications can be huge."

We see a similar effect in other areas, too. The power grid has such strong interdependencies that a single minor outage can propagate across an entire continent, pulling services down on its way. As we saw in the first chapter of this book, meanwhile, the international food trade web is so intricate that a local shortage immediately creates a global problem. Only traditional farmers who use their own seed and manure are unaffected.

"Once you acknowledge that your data are full of uncertainties, planning becomes a lot simpler," De Kok promises. "A lot of information can stay local; checkout data can comfortably remain at the relevant branch. That's a totally different principle to the way we currently develop technology. Companies needn't invest millions in a vain attempt to pin down everything going on within and beyond their walls. Rather than endless data collection, we should accept that some things are unpredictable and uncertain. In some cases, we do know how transport times, processing stages, and purchasing behavior can vary, and some factors fluctuate more strongly than others. We need planning software that takes greater account of uncertainty. It remains difficult to take full account of statistical variations like this, but that's the challenge to be met in what is a new field of research."

Ton de Kok is not alone with his critique. In recent years, complexity research has come up with totally new approaches for handling uncertainty in planning processes. Although it's still early, similar thinking about decentralization can also be detected in other areas that are characterized by complex networks and local factors, such as telecommunications, power networks, and computers. Modern telecom techniques use decentralized "distributed intelligence" to respond rapidly to network blockages.

Perhaps there's a lot we could learn from nature. A great deal of decentralized regulation occurs in our own bodies, too, which can cope with shifting local needs while still continuing to function as a unit. Other approaches will no doubt arise as the research proceeds. But if Ton de Kok has anything to do with it, the future lies in decentralized planning. Combining the concept of decentralized planning with the concept of stochastic models and generic insights drawn from them yields more effective network planning and control mechanisms that are totally different from what is used today.

3.7

ADVANCED MACHINES

What looks like a cake tin on wheels is working its way around the room. The robot vacuum cleaner is just as noisy as a normal one, but there's an important difference: You don't have to lift a finger to clean your floor. The dirt collects inside the robot's body, swept up industriously by the rotating brushes and sucked in by the motor. The machine's sensor flickers over the spot where some bread crumbs just fell, telling it that this is an especially dirty place, which requires an extra sweep for good measure. At the edge of the stairs, the cake tin detects the drop and changes course in the nick of time. Having surveyed the room three times, the robot concludes that its job is done. Everything is clean. No more arguing over who has to vacuum the floor. Let the machine do the work while you sit in a comfortable chair, maybe with another robot for a pet. You can already buy devices like this for a couple of hundred dollars.

In fact, much of the Industrial Revolution is about machines working for us. That has dramatically changed productivity and labor. In our households, too, we have a number of machines that do the work for us. Examples are our washing machines and dryers. But for as long as machines have existed, we have dreamed of robots that could take over more tedious chores—metal people who would obey our every order and do our work for us—open the door, boil the potatoes, fix the car. It's no coincidence that *robot* derives from the word for "work" in most of the Slavic languages. Robots spark fantasies of large factories full of metal workers lifting boxes, toiling on the production line, and designing new products at their drawing boards. These are some serious toys. They extend our human capacities in much the same way as all the other tools we have developed in the course of our history. Some are already in use in our daily lives, including ones that make independent and crucial decisions without seeking our input. The antilock brakes in our cars, for instance, is a kind of robot, too. It acts faster and more precisely than we ever could if the car suddenly threatens to skid out of control. In practice, though, we tend to label as "robots" only those machines that are intended to mimic some kind of human behavior. These could now offer some of the tools we need to tackle the major problems of our era. We're no longer simply talking about vacuuming or welding but about helping to care for old people, improving the precision of surgical operations, and streamlining our transport systems. Robots will also feature more prominently in our daily lives as human–machine interfaces steadily improve in response to advances in voice recognition.

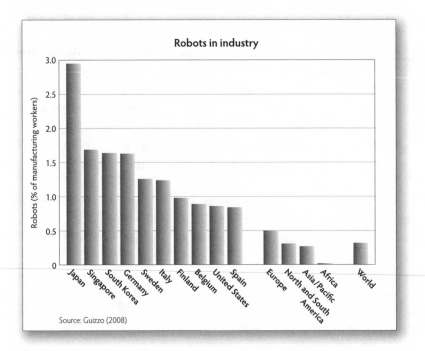

Source: Guizzo (2008)

Robots are monomaniacs that can repeat the same action thousands of times, often with greater precision and speed than humans can manage. Much of the Industrial Revolution is about machines working for us. That has dramatically changed productivity and labor. But industrial robots are specialist machines, not versatile employees capable of performing different tasks every day. To be even more useful, robots should be more felxible in their capabilities. Industry increasingly demands flexibility. *Source*: Guizzo, E. (2008). The rise of the machines. *IEEE Spectrum*, 45(12), 88.

HUMANOIDS

Engineers are getting better and better at emulating human characteristics in robots. There is a ceaseless competition to design robots capable of replicating human capabilities as closely as possible. Honda took the lead at the beginning of this century with Asimo, which resembles a midget in a space suit. Asimo is a toy human that walks upright. Its arms and legs are controlled mechanically beneath a snow-white plastic skin. Cameras behind the visor of its space helmet survey the robot's surroundings. Asimo can already tighten screws, fetch newspapers, and carry messages, effortlessly avoiding chairs and tables in the process. Honda even used Asimo as a receptionist at one of its offices, greeting visitors and bringing them cups of coffee. Honda's toy people have proved a great success in commercials, at conferences, and in playgrounds. Asian businesses have invested billions in all-round humanoid robots like this. New techniques are being explored that can later be applied to industrial robots. More recent examples include Hoap (Fujitsu, 2001), Qrio (Sony, 2003), Wakamaru (Mitsubishi, 2005), iCat (Philips, 2005), HRP-4C (AIST, Japan, 2009), and many more. These robots imitate a whole range of human characteristics, feeding themselves with electricity, learning, moving like humans, and communicating with us.

Coordinating the movements of humanoid robots is an intricate task. Robots already exist that can climb stairs, jump, and dance. It's all still very mechanical, but it's fun to watch. Robot movements are still far removed from real human motion. Humanoid robots aren't terribly good at walking on two legs. That's hardly surprising, as it takes humans themselves at least a year to get the knack. When we sense that we're losing our balance, we throw out our arms or shift our hips to regain a firmer stance. Natural limbs have all sorts of subtle capabilities for adjusting their movement. The human foot alone is made up of numerous little bones and muscles that enable us to walk easily across bumpy or soft terrain. The muscles tense one after the other so that we can lift our feet in a fluid movement. Try building a machine capable of doing that. Anyone who has an artificial knee or hip will be familiar with the stiff motions they produce; artificial limbs rotate less smoothly than real ones whether in human beings or robots. Robot joints are less flexible than ours, so the wear on them is a significant maintenance issue for industrial applications.

Some designers have produced complex metal joints surrounded by tiny motors that imitate human muscles. Other robots have big flat feet. Those keep them from falling over but also mean that the robots in question could never hope to walk along a bumpy path in a forest or field. The question is, however, whether the human model is the most efficient one. Nobody would object to a doglike robot to tighten screws, provided it was stabler.

Robots have already become sufficiently human, however, to compete in the RoboCup—the annual cybernetic world soccer championships. The 2009 tournament saw robots doing throw-ins thanks to improved control of arm movements. The metal soccer players can now create mental models—scenarios of the different ways in which the game could develop. After each tournament, the engineers share their secrets, enabling opponents to learn from one another. This significantly speeds up the rate of progress—not to mention ensuring rapid jumps and slides in the league table. What will really make everyone sit up and take note, however, is when robot soccer players become better than humans. The RoboCup organization has set itself the goal of producing robots that can beat a national soccer team by 2050. The ultimate objective is not to revolutionize soccer but to perfect robots in ways that enable them to perform other tasks—in our homes, offices, and factories, for instance.

NOT JUST ANOTHER TOY

It's fun when robots deliver a newspaper or play soccer, but it's hardly going to save humanity. Nevertheless, copying humans and developing robotic soccer have a clear appeal. The practice they offer will help produce machines that are useful and which give us the tools we need to push humanity forward. Serious robots are often less human. Most of them don't look like metal people any more than the robotic vacuum cleaner does. Welding, painting, and assembly robots have been toiling in our factories for more than 40 years now. They reduce labor costs and offer mass production combined with a high degree of precision. It all began with a simple assembly robot at General Motors in 1961, but subsequent development was rapid. If you take a look inside a modern car manufacturing plant, you'll find one robot for every ten human workers. According to the International

Federation of Robotics (IFR),[1] something like a million industrial robots are now employed around the world. Future industrial robots will be able to perform more tasks, respond to changes more rapidly, and detect imminent failures.

There is a significant difference between these machines and people. Industrial robots always do exactly the same thing, just as they have been programmed. Once you've taught them how to weld or paint, they can repeat that same action thousands of times, often with greater precision than we humans can manage. They are monomaniacs, and that's precisely their strength. In many cases, the reason for introducing robots is often not about cost but precision. A robot is, after all, more expensive in most cases than a salaried person.

Being a monomaniac has its drawback, too. If the model of car changes, robots have to be taught what to do all over again. Industrial robots are specialist machines, not versatile employees capable of performing different tasks every day. To be more useful, robots should be more human, maybe not in their appearance but in terms of their performance. Industry increasingly demands flexibility. Having welded a steel shelf unit together, the robot should ideally be able to switch straight away to a steel cabinet. That's where the knowledge gained from building toy humans comes in. Playing with toy robots teaches us how to respond faster to changes and how to detect imminent errors. Industrial robots need to be more human in the way they perceive their surroundings and how they respond to unexpected situations. For this reason, collaboration between machines and interaction with humans are becoming increasingly important.

OPERATING THEATER

"Robots and people make for a pretty formidable combination," observes Maarten Steinbuch, professor of systems and control at Eindhoven University of Technology in the Netherlands. He has a particular interest in the way devices are controlled—an area in which he admits a number of breakthroughs are needed. "Human beings and robots sometimes work extremely closely together. One example is abdominal operations in which the surgeon uses a joystick to control a robot." In this instance, the robot arm is a tube that enters the abdomen through a small incision. A camera and cutting instruments are attached to it. The surgeon controls the robot from a console that provides an enlarged, three-dimensional image, ensuring greater precision. "Surgeons can zoom in on the details they want to see, and their movements are enlarged accordingly," Steinbuch explains. "They can then operate on a gall bladder, remove the appendix, or repair a hernia."

The current generation of medical robots also has its limitations, however. They are large and heavy, making them difficult to deploy. What's more, they take theater staff too long to set up. "But robots are becoming lighter and easier to use," Steinbuch says. "We're also working on medical robots that are more responsive to the surgeon." This picks up on the fact that surgeons can't "feel" the patient when they work with controls on a console. The tactile element is, however, an important surgical skill, as surgeons learn how to evaluate the tissue using their fingers. "It's extremely difficult to provide the feedback that will give surgeons back that tactile sense. You have to design the medical instruments in a different way if you want to detect the forces that are involved. We're carrying out research into perception, too, so that we can learn precisely what surgeons want to feel and how. The electronic controls have to transmit changes in the tissue to the surgeon."

In some cases, that sense of force will be something surgeons are able to experience for the first time. "That's the case with eye operations, for instance, which is one of the applications for which we're developing a medical robot," Steinbuch continues. "Up to now, surgeons have never had any physical sensation when working directly on the eye, as the tissue is extremely soft. We can give them a kind of tactile feedback they've never had before." The techniques being developed for these robots have many other applications. "We're studying the possibility, for instance, of developing remotely controlled catheters for heart operations. And our tactile feedback approach is also being applied to the control of vehicles."

PERSONAL ROBOTS

Robots will feature more and more frequently in our daily lives, Maarten Steinbuch thinks. The population is aging, and there are already too few people to help the elderly to dress, wash, and eat.

There'll be a serious need in the future for artificial hands to help around the home. Toyota believes that care robots will soon be as common as hearing aids. Although these things aren't too far away, engineers still have a number of problems to solve. "Easing a support stocking up a swollen leg takes the kind of precision and flexibility that right now only a human being can provide. It's way too complex for robots. And robots are confronted in the home with unforeseen situations to which they have to respond appropriately. Robots haven't sufficiently mastered observation, image processing, or movement to be able to do that yet. The difficulty is that robot design requires you to combine knowledge from highly divergent disciplines. It involves mechanical engineers, electrical engineers, physicists, and IT specialists, each group with its own methods and its own tools. All of which you have to cram into a single high-tech machine."

A number of breakthroughs are also required in the way devices are powered and controlled. "Limited battery capacity remains a serious obstacle at this point," Steinbuch admits. "There are lots of functions that will require robots to be able to move around independently for longer." Meanwhile, we may teach the robots to find wall sockets to recharge their batteries. Sensors are nowhere near rapid enough yet either. "You need to observe your surroundings faster if you're going to respond effectively. The sensor technology isn't up to that yet." There's a lack of standardization, too, which means each new robot project has to start from scratch. Some kind of common development platform is urgently needed so that previous developments can be readily incorporated into new designs. That would also facilitate the development of operating software for robots, which remains very inefficient at this stage. "Robot software is complex," Steinbuch confirms. "Tracing and correcting errors are difficult and time-consuming. Programming is hard—especially when different robots have to collaborate or when human beings come into the equation. Robots have to pass components between themselves and coordinate their actions. But they can also get in each other's way. The more flexible the machines become and the more they work together, the harder it is to keep track of whether everything is going smoothly."

These difficulties aren't surprising, given that every robot movement is centrally controlled and determined by a series of rules. An alternative is to make robots more adaptive. When they measure the effect of every movement they make, they may learn how to shift their legs or move their arms at just the right moment. That turns out to be much easier than trying to calculate every last movement of a series of electric motors. Robots of this kind can even walk on stilts. This approach has arisen from our knowledge of complex systems. And it's merely one of the potential alternative approaches. Whether robots are playing soccer, lifting an elderly woman out of bed, or operating on your eye, a great deal of interaction is always required. Robots are being asked to perform progressively more complex tasks in which feedback and interconnection are increasingly important. Greater knowledge of complex systems will contribute substantially to their improvement.

"We're only at the beginning right now," Maarten Steinbuch thinks. "Robots represent the next industrial revolution. The time is riper for them now than it was 20 years ago. Some years from now, there will be some kind of robot in many houses. The technology is beginning to converge. We're learning how to manage the complexity of robot interaction with people.

Part 4
HUMANS

4.0

THE NURSERY OF LIFE

Human beings are much more complex than any technology we could devise today. How many machines are good for 80 or 90 years of service? Our immune system—set up at birth—is able to repel diseases that don't even exist yet. Most viruses that proliferate 50 years after we were born can be defeated just as easily as maladies that have been dogging humans for generations. Effective health care means that—in most regions of the planet—we are living longer and longer. All the same, human beings are not perfect: We get sick and we wear out over time. In the wealthier regions, we spend a great deal of money trying to get as close as possible to a 100-year span. Our greatest task is to bring a long and healthy life within the reach of as many people as possible. New technology is required to hold down the cost of health care, to nip outbreaks of disease in the bud, and to ease discomfort in our old age. Scientists believe that substantial benefits can be gained by identifying abnormalities earlier. A cancerous growth measuring just a few millimeters is still relatively harmless, and an infection caught in its early stages won't leave any scars. Although techniques for accurately diagnosing incipient abnormalities can often be very expensive, prompt diagnosis generally means that treatment will be easier, cheaper, and more likely to succeed. Thus, we can end up saving money despite the need for expensive equipment.

To adequately fight the outbreak of diseases in the future, our technology must be able to respond more rapidly. This could pose a particular challenge because there is also a trend at present toward superspecialization, which is fragmenting medical knowledge and slowing down responses. Take the science of ophthalmology in which the various specializations focus on extremely specific parts of the eye. This is fine once a precise diagnosis has been made, but it could be a significant problem if the patient consults the wrong doctor at the outset.

The way we currently approach diagnosis needs to change. A patient presenting today with an ailment that can't be readily identified will be referred to a specialist and will have to make a series of appointments for all kinds of blood tests, scans, and other checks. The process could be significantly improved and accelerated if these examinations were combined. Equipment already exists that can carry out several types of scans at the same time, and scientists dream of a universal diagnostic device capable of scanning everything in a single pass.

In 20 years, that dream will gradually move toward reality. This will also blur the distinction between diagnosis and therapy, because future scanning equipment will also be able to start treatment immediately. Work is going on in laboratories to develop smart pills, powders, and drinks containing agents that will independently seek out diseased parts of the body to neutralize malfunctioning cells, using the energy supplied by the scanning devices. Scanning and therapeutic devices will thus be integrated. Waiting times will be reduced because patients will be treated immediately after diagnosis (chapter 4.1).

To treat ailments more effectively, we also need a much clearer understanding of what really makes us ill. That ultimately comes down to a deeper insight into microbiology—the chemistry and physics of our bodies. Any malfunction in our body's system confronts us with a complex mass of different processes that we are only now beginning to understand. The challenge this poses is another complex problem. Practical solutions will take the very best of our knowledge in fields like electronics, mechanical engineering, telecommunications, and control systems, as well as the medical sciences (chapter 4.2).

Infectious diseases are one of the major health concerns of the twenty-first century. Even if the pandemic of 2008 was less devastating than feared, a new influenza pandemic could be both imminent and catastrophic. Any such pandemic could kill hundreds of millions of people, lead to a partial die-off of the global population. Managing it would be difficult because supplies of antiviral agents are limited and expensive. People who live in countries without vaccine companies—more than 85 percent of humankind—will have little prospect of immunization. New approaches are therefore needed to confront an imminent pandemic (chapter 4.3).

There's more to health care, however, than simply detecting and dealing with abnormalities and suspect places in the body. The key attitude we need to pursue in the future is probably a thorough knowledge of how to prevent problems in the first place. It's also a question of keeping active and able to fend for ourselves. Elderly people in particular can benefit from equipment that helps them remain independent longer. Technology can do much to make old age more pleasurable. We might not live any longer, but we can at least make the final stages of life more comfortable (chapter 4.4).

There are many things we do not yet understand. Our bodies—and nature as a whole—still hold many secrets. Nature was coming up with smart solutions a billion years before humanity made its appearance, and there's much we can learn from it. Evolution has optimized our ability to ensure the future of our species. It pays to imitate the natural world and, where appropriate, to improve on its processes.

4.1

THE TRANSPARENT BODY

The "easy" diseases have pretty much been beaten in the Western world, leaving doctors to contend with the more complex illnesses that stealthily overrun the body. Two-thirds of the deaths in the United States are now attributable to cancer or coronary disease.[1] By the time these conditions manifest themselves, it's often too late to intervene.

Treatment is only likely to succeed if early signs of cancerous growth or clogging arteries can be detected. A tumor measuring a few millimeters across is plainly less threatening than one the size of a tennis ball, not least because there is less risk of metastasis at an early stage. The focus is therefore on enhancing rapid diagnosis, which in turn means improving medical imaging.

Eighty percent of all diagnoses are based on images. Yet many small but life-threatening physical processes are still missed by the scanners, echographs, and other devices that peer inside our bodies. Growths measuring less than a centimeter tend to be overlooked, so scientists are constantly working on techniques capable of offering a more detailed internal picture. Breakthroughs in imaging technology can mean the difference between life and death. They'll enable us to intervene sooner, boosting the patient's survival chances.

Little more than a generation ago, X-rays were the only means we had of looking inside the human body. The images they produce are flat, however, and lacking in depth information, which can make them hard to interpret. An ingenious technique was therefore devised in the 1970s that allowed a single three-dimensional image to be created by combining a series of X-ray photographs. The CT (computerized tomography) scan was the first technique to produce a genuine three-dimensional image of our insides. Doctors could now tell, for instance, whether an abnormality was located on top of a bone or beneath it. Several other techniques for producing three-dimensional images of the body have since become available, some of which require patients to be injected with a contrast agent to highlight specific parts of the body.

In the case of a PET (positron-emission tomography) scan, the patient is injected with a substance that closely resembles natural sugars, with the difference that the sugar molecules have been rendered slightly radioactive so that they can be detected externally. The imaging scanner, acting like a camera, can then track precisely where the sugars go—namely, to tissues that require a considerable amount of energy. Hence, PET does not generate images of our anatomy as such but of energy consumption at different locations within the body. That makes them especially useful for detecting metastases, which consume much more energy than other tissues.

The quest is on to find new contrast agents capable of revealing more processes and finer details of the human body. This has led to the development of a separate branch of science called *molecular imaging*, which focuses on new agents that resemble physiologically active molecules and attach themselves to abnormalities and "active" locations in the body. Agents are being researched, for instance, that latch onto moribund cells, providing clues as to whether a cardiac muscle is close to giving out. Mapping these molecules should make it possible to detect heart attacks early enough to save the failing muscle. Scientists are also working on a new generation of binder molecules that will bring early detection of individual cancer cells within reach. This will in turn allow intervention before irreparable damage occurs. The new contrast agents aren't necessarily radioactive; the more sophisticated substances can be observed from outside the body thanks to their magnetic or optical properties. There are agents, for example, that emit light when you shine a laser on them.[2]

FEWER SCANNERS, MORE DATA

Molecular biochemistry is continuously refining the available scanning technology, although it will take time for these advances to find their way into standard hospital equipment. We can look forward to a steady flow of new examination techniques, each of which will deliver images of a different aspect of the body. This doesn't necessarily mean that the number of scanning devices will also increase; in a parallel trend, these individual techniques are being combined in the same device. We already have scanners, for example, that can perform simultaneous CT and PET scans. The PET image provides functional information about locations in the body that are consuming a lot of energy, while the CT scan offers a high level of structural detail.[3]

The ideal scanner would display everything down to the last cubic millimeter, detecting insidious conditions like cancer as soon as they begin to develop. New scanner technology like this would mean a huge step forward in combating two-thirds of the Western world's fatal diseases, at least in theory. All those images would still need to be analyzed, and that's a growing problem. The average institution already generates thousands of gigabytes of images a year, and smart scanning techniques will only multiply the amount of data. Combining different scanning techniques will generate a flood of additional information about the body.

We discussed these challenges with Jacques Souquet, an expert in medical imaging who spent many years at Philips Medical Systems and who has several ultrasound patents to his name. Having returned to his native France, Souquet set up the medical equipment firm SuperSonic and now also heads SonoSite. Both companies continue to develop more advanced forms of imaging and therapeutic devices. "Medical imaging specialists nowadays have to view over 10,000 images a day in order to do their job properly. We're reaching the point where it will simply become too much," he warns. It's time-consuming and expensive to study all this material in detail, and the limits of doing this work manually—or rather, visually—are rapidly being reached. There's no point in improving scanning techniques, Souquet believes, if the result is to put physicians under even greater pressure. "We've already exceeded our human capacity to evaluate this material."

It isn't only new scanning techniques that are adding to the stream of data. Improvements in existing technologies are doing so as well. Each individual scanning device is pouring out ever more data. And the number of scanners is steadily increasing as their price and size decrease.

Scanners will also be used in new contexts, according to Souquet. "We've developed an ultrasound imaging unit for the army that can be connected to a handheld computer. It's so small you can use it in a whole new way. The device resembles a stethoscope but produces a visual representation of what's going on inside the body. Small, easy-to-use, and fast imaging equipment will start to appear in hospitals, too. Expect an increasing number of scanners in consulting rooms, hospital wards, and operating theaters. Specialists will soon be able to carry diagnostic equipment around in their pockets, just as they've always done with the conventional stethoscope." The same technology will increasingly find its way into our homes, too. Electronic systems that detect heart failures are already available, and portable medical diagnostic systems for measuring blood glucose levels or oxygen content are also common. A wide range of other measurements will soon be added. The equipment is becoming smaller, easier to use, and more sensitive. It won't be long before we'll be able to detect cardiac arrhythmia at home as well or heart attacks, for that matter.

That's the future according to Souquet: "People are already having their heart function monitored remotely, and miniaturization will make that a lot easier. You'll be able to carry an electrode around in your wallet or mobile phone that will pick up and transmit your heart rate. Services like this were created for people with a heart condition. But the biggest users are perfectly healthy baby boomers who are scared of getting ill." Souquet is concerned about the implications of this development, which means medical data from healthy people will also start to flood into hospitals. "Doctors often have a suspicion that something's not quite right. Then what? Not every condition is as clear-cut as a heart attack. How are you supposed to react to a heart murmur, a variation in lung capacity, or slightly increased cholesterol in your blood? Is it a sign that something serious is about to happen? Doctors are going to be confronted with lots of worried patients with whom there's virtually nothing wrong. But to be on the safe side, extra examinations will be necessary. That not only increases the cost of health care, but it will also generate yet another wave of imaging data."

THE COMPUTER TAKES OVER

One way of coping with the flood of images would be to improve visualization. All the big suppliers of medical imaging equipment are doing serious work in this field. The trick is to combine data from different sources into a single, easy-to-navigate rendition. This prompts producers of medical software to look to the games industry to learn how to navigate through images of the body using a mouse or joystick, like gamers exploring a fantasy landscape. Each organ is colored differently and appears to float in space. The computer hides irrelevant areas and adds detail whenever required. It is now possible to take a virtual tour of the body. We can travel down the windpipe, for example, and into the lungs, zooming in on a diseased pulmonary lobe to identify the seat of an infection. Doctors will be able to explore blood vessels, urinary passages, and the digestive system in a similar way. Intensive work is also going on to bring advanced visualization into the operating theater, where it will combine with the latest scanning techniques to enable surgeons to see right through the patient's body. They will be able to monitor with great precision, for instance, as a needle is inserted through the spine and into the spinal fluid. The patient will be rendered literally transparent.

Improved visualization will make it easier to navigate through the body. It won't be enough, however, to detect the beginnings of every possible illness, Jacques Souquet warns. "As the resolution of imaging techniques rises, the complexity of the images themselves will increase enormously, and the number of relevant details will explode. We're going to have to build more intelligence into the software so that it can interpret what it sees. The computer has to help diagnose by, say, drawing a red circle around any abnormalities." To achieve this, knowledge that currently resides in doctors' heads will have to be captured in the computer. The software must learn the many different factors a radiologist evaluates when examining an image—darker patches in the gut, hardening in the kidneys, or the dimensions of the heart, for instance. The computer must learn how to select the right details from this immense body of information.

One way of doing this is to use the large image archives that have already been built up at hospitals. The computer can be fed old patient data and trained to recognize the characteristics of different conditions. This results in complex statistics, as certain symptoms occur with several diseases, and some conditions can be accompanied by a wide range of different symptoms. Armed with these statistics, the computer program can then analyze the symptoms of new patients and suggest possible diagnoses. The more data that are available about the patient—from blood tests or ECGs, for example—the more the computer will be able to eliminate particular possibilities.

The software could also suggest which tests would provide greater certainty.

Analysis of this kind will become easier as we are scanned more frequently. The computer will then identify anything new that appears in the images, allowing the experts to focus on that particular location. Similar techniques are already well established in astronomy, which has long been confronted with challenges like this, prompting astronomers to develop refined analysis techniques that highlight changes in successive images. These can also be applied to the human body. Computers may already interpret mammograms more effectively than doctors can. Indications are that a computer outperforms the human professionals in the diagnosis of breast cancer.[4]

But that's just the beginning. Computer assistance will become increasingly important as medical knowledge continues to compartmentalize. Heightened specialization—and even superspecialization—might be necessary as our knowledge of diseases deepens. Yet specialization is also increasingly problematic because patients risk being referred to the wrong specialist at the beginning of the diagnostic process. Computers have the potential to provide medical professionals with knowledge that would normally lie outside their specialist area.

The role of the doctor is bound to change as computerization continues. We should expect computers to start producing better diagnoses than doctors for more and more conditions. And possibly, machine diagnosis will eventually be preferable in the case of certain diseases. Will we still need a human being to tell us that we are sick? Jacques Souquet stresses that it is the doctor's job to diagnose, not the machine's. "But software will increasingly draw the physician's attention to possible abnormalities so that the doctor can make the final judgment. But if the flood of data generated by ever more detailed and varied medical images is to be managed, it is surely inevitable that the next step will be for computers to take over actual diagnosis, too."

FROM SCANNER TO THERAPY

The degree of computerization is bound to become even greater as treatment techniques are incorporated directly into the diagnostic process. As a developer of diagnostic equipment, Jacques Souquet finds himself focusing more and more on therapeutics as diagnosis and treatment increasingly overlap. He dreams of equipment capable of doing both: internally scanning the patient and then commencing treatment immediately. The key is to use scanners in combination with contrast agents, which can attach themselves to sites of disease with growing precision. The next step is obvious, Souquet says: "If contrast agents are going to zero in on the affected area, they can also initiate treatment as soon as they get there. Chemists and biologists are working on techniques to add medicines to them, using particles that will act as a kind of nanoscale capsule. The patient will be injected with a solution containing these particles before getting into the scanner. The combination of contrast agent and medicine will travel through the body and attach itself to affected areas like a tumor or diseased cardiac muscle. The nanocapsule will then open, releasing the medicine."

It will be possible to activate the payload externally. "We're working on techniques using ultrasound to vibrate particles that have latched onto a tumor," Souquet confirms. The vibrations cause them to break apart and release the medication. Any other approach in which energy is administered from outside could work, too. Precise treatment of this kind will open up new ways to combat diseases like cancer. At the end of the day, using selective agents is a great deal more efficient than the chemotherapy we have now. The medication currently administered passes throughout the whole body—hence, the widespread side effects. It would be a huge advance if the drugs were only released at the necessary location."

The new approach will combine diagnosis and therapy in a single device, with major benefits in that you will no longer have to wait for treatment once your condition has been diagnosed. Healing can begin during scanning itself. "That will steadily narrow the boundary between diagnosis and therapy," Souquet maintains. It will also increase complexity as elements of the medical process that were hitherto loosely coupled become strongly intertwined. All relevant data will have to be processed and analyzed in near real time. The steady automation this will require will increasingly take not only diagnosis but also therapy out of the hands of physicians.

4.2

PERSONAL MEDICINE

No two individuals are alike. Some people are genetically predisposed to develop asthma, whereas others can cheerfully live a hundred meters from a major highway with no adverse effects. Genetic predisposition also plays an important part in the efficacy of drugs and in the progress of diseases like cancer, heart failure, and diabetes.

Individual differences make doctors' work more difficult. They can never be sure precisely how susceptible a person is to a specific disease or how effective a particular medicine will be. We can measure all sorts of things, but what do we have to know before we can accurately predict whether a given person will fall ill? Part of the answer is hidden in our genome: Inherited defects and sensitivity to medication show up in our DNA.

The map of the human genome was colored in at record speed at the beginning of this century by two rival research teams, which ended up publishing their results simultaneously in 2001.[1] Their achievement was compared with the first moon landing and the invention of the wheel. One of the competing groups was headed by American Craig Venter, who continues to spread the DNA gospel enthusiastically. Initially, Venter was part of the U.S. government–sponsored Human Genome Project, but he left the group. He founded a private company to create a database of genomic data. Venter characteristically mapped his own DNA, revealing that he bears a heightened risk of alcoholism, coronary artery disease, obesity, Alzheimer's disease, antisocial behavior, and conduct disorder. Unfazed, he enthusiastically published his complete genome on the Internet. "A lot of people are scared to have their DNA examined," he says. "They think all their inner secrets will be revealed. Even medical students are wary about supplying their DNA. But the course of our lives isn't genetically determined, apart from exceptional cases where life expectancy is reduced by a serious hereditary condition." Most people aren't aware of the subtle mechanisms of genetics, he adds. "People think like 1980s scientists. Possibilities for analyzing DNA were limited back then. All you could do was link certain diseases to defective genes—Huntington's disease and cystic fibrosis, for instance. Most people still have that picture. They think that every human characteristic is determined by a single gene." Hence, there's a great deal of talk about gene passports mapping all an individual's genes and revealing his or her destiny. "But it's not that deterministic," Venter stresses. "It turned out later that it's only in rare cases that a change in a single gene causes a disease. It's usually much more complex than that. Hereditary predisposition for cancer is more often caused by a complex interplay between many different genes. It's sheer chance that determines whether or not you develop cancer. A defect in one gene merely increases that possibility. But that notion didn't reach the public."

Venter gives the example of colon cancer, the development and suppression of which involve at least thirty-four different genes.[2] "We happen to know of one gene that suppresses cancer of the colon: It triggers the creation of enzymes that eliminate the cancerous cells. If you have a defect in that gene, you have an increased probability of developing colon tumors by midlife.[3] But that gene is only one link in a chain. So if you do turn out to be predisposed to cancer, it doesn't mean you're predestined for it. The opposite is equally unclear: If the gene is good, that still doesn't mean you're totally risk-free. Measuring the gene merely tells you something about probability; it's pure statistics. That's how we'll use genetic medicine. If you know you have a heightened risk of developing colon cancer, you won't wait until you're 50 to have your first intestinal examination. It's much more cost effective to carry out a colonoscopy at an earlier age; if you find anything, you can start treatment immediately. An operation performed in the early stages of colon cancer is much less expensive, less invasive, and more likely to succeed."

Genetic patterns can also be useful in fine-tuning treatment. When the first symptoms manifest themselves, genetic information can help identify the best type of medication, the most effective dose, and even the likelihood of survival. For example, new patients with breast cancer can be given an individual prognosis of how successful their treatment is likely to be based on anomalies in the BRCA2 gene. If you could draw on genetic information, you'd be able to calculate someone's personal medication and dose. For many diseases, this still

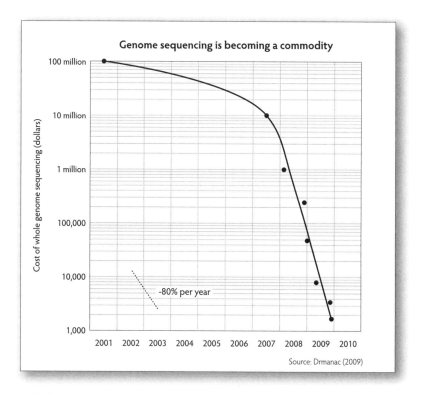

Genome sequencing is becoming a commodity

Cost of whole genome sequencing (dollars)

-80% per year

2001 2002 2003 2004 2005 2006 2007 2008 2009 2010

Source: Drmanac (2009)

Medical specialists will soon be able to sequence patients' genomes as a matter of routine. This will open up new possibilities for personalized treatment that takes into account the individual's specific inherited characteristics. Using that information efficiently is a challenge that will require progress in computational science. *Source*: Drmanac, R., Sparks, A. B., Callow, M. J., Halpern, A. L., Burns, N. L., et al. (2010) Human genome sequencing using unchained base reads on self-assembling DNA nanoarrays. *Science,* 327 (5961), pp. 78–81.

remains a dream. Yet a deeper understanding of the complex interplay of genes and proteins in our bodies should allow more personalized diagnosis and treatment.

UNRAVELING CONTROL MECHANISMS

The race to map the human genome was followed by the emergence of a new generation of geneticists seeking to unravel the mechanisms by which DNA regulates cellular processes. The primary focus today is no longer on the "geometry" but on the way these processes function and malfunction. DNA can't be understood without understanding the thousands of different proteins it creates, each of which has its own task to perform in the cell.

Regulation of the complex network of interacting proteins is influenced not only by the structure of the DNA but also by its relative accessibility. DNA is folded, and as a result, not all of the genetic code is active in the body. So the way it's folded is relevant. And there are many other factors that determine how the genetic information in the DNA is used. There are switches that turn functions on or off. There are triggers from outside that provoke a reaction. And above all, there are intricate combinations of all these factors that orchestrate the actions of proteins.

The cell's control network reveals fresh levels of complexity with each new discovery, further complicating the web of interactions between DNA and proteins. Until recently, for instance, an individual gene was thought to contain the blueprint for one specific protein designed to perform one specific task within the cell. Recent research shows, however, that a single gene may assemble a wide variety of different proteins, switching its mode of operation in response to input from its surroundings. That means the same gene can regulate different cellular functions by expressing different proteins. More than 70 percent of the genes making up human DNA display this versatility, with some even capable of producing thousands of different proteins.[4]

Despite all this complexity, the DNA protein network turns out to be incredibly stable. Many proteins play only a minor role in regulating the cell. The network would retain its stability even if a few types of protein were missing.[5] We now know that certain other protein types can partially take over the lost functions, enabling the organism as a whole to survive.

The stability of cellular regulation is also apparent in the natural variations that occur within a species. This was another groundbreaking result obtained by Craig Venter, this time in 2007 when he was the first to sequence both helices of a human DNA strand (once again, his own). One helix contains the information inherited from his mother, and the other information from his father. Analysis of both parts of Venter's DNA revealed that at least 44 percent of the genes from Venter's mother differ from those from his father. Venter's team identified 4.1 million variations among 2.8 billion base pairs—at least five times more than previously thought.[6] If this is a true reflection of *Homo sapiens'* variability, then our cellular regulatory mechanisms are astonishingly stable. You'd certainly notice the difference, for instance, if you reset 44 percent of the switches in the control room of a power plant.

The stability of cellular control comes at a price. Certain proteins play a pivotal stabilizing role as they connect many different processes and maintain the overall balance. Without them, the cell's regulatory network would be fragmented. This knowledge might be exploited in cancer treatments. The customary robustness of the regulatory network makes it hard to kill cancerous cells with drugs that merely disrupt cellular controls. However, a drug that targets this Achilles heel of cellular regulation with sufficient force might be able to do so.

We are only now beginning to understand the stabilizing mechanisms at work within cells. If we could discover how the system might go awry, triggering cancer or another disease, we would be better able to predict a patient's susceptibility to different illnesses. This might also provide clues to new medicines that target the cells' weaknesses. We could learn—by analogy with other nonlinear dynamic networks—to slow down deviations before the regulatory network becomes unstable.

These insights are the fruits of an immense amount of research by many scientists following the initial mapping of the human genome by Craig Venter and his rivals. Further progress will depend on the skills of the mathematicians who must now work their way through gigabytes of genetic data. Some of this comes from medical archives that we also encountered in the previous chapter. Genetic material from cancer patients has been stored systematically and on a large scale for decades now together with details of the progress of their illness and the efficacy of the treatment. The limiting factor in research of this kind is the enormous processing power needed to combine all the historical information with the relevant molecular data. Achieving that processing power will be a major technical challenge but one that could provide the breakthrough in integrating this huge body of genetic information and case histories.

Another limiting factor at this point is that a great deal of research relies on DNA sequencing, which remains a costly procedure. A lot more DNA data will be needed before we can unravel the regulatory system of a cell. Venter predicts that the cost of DNA analysis will fall quickly and that DNA sequencing will become increasingly straightforward. "It took me 9 months to map my own genome. If we really want to make medical use of genetic data, we're going to have to be able to sequence a human genome in a matter of hours or even minutes and for less than $1 thousand a time. At that kind of price, we'll be able to examine lots of patients and then build databases of medical information. That will generate a mass of significant data for scientific research, which will further accelerate progress."

RE-CREATING LIFE

Meanwhile, Craig Venter has embarked on a new project. He wants to create new DNA with a view to reprogramming life. "In every glass of seawater or liter of air that we study, we find new bacteria and viruses with new properties. We can apply that knowledge to make new forms of life. In the same way that you can use transistors, resistors, and condensers to make every conceivable type of electronic circuit, with the right combination of bacterial properties, you'll be able to carry out any chemical process. How fantastic would it be if you had an organism that could extract CO_2 from the atmosphere and convert it into polymers? Or a bacterium that uses sunlight to extract hydrogen from water? The genetic world offers millions of building blocks you can use to construct properties like that."

In 2008, Venter achieved a new milestone in the race to uncover the secrets of life when he synthesized a complete genome from scratch for the first time. He strung molecules together to produce an exact copy of the DNA of a bacterium called *Mycoplasma genitalium*, which causes gonorrhea-like symptoms in human beings. At that time, it was the species with the smallest number of genes then known to sustain life. That meant Venter had to connect "only" 582,970 base pairs to copy its 521 genes. As a precautionary measure, he omitted the bacterium's pathogenic gene.

"The next step was to build DNA into a cell," Venter says. "It is a little like changing a computer's operating system—replacing Windows with Mac software. In that way, we'll gradually move from reading genetic information to being able to write and implement it." This success in May 2010 was hailed as the first artificial life, but

Venter stresses that this wouldn't amount to creating life out of nothing. "I only replicate gene patterns that we know; I don't create novel genes. But we'll ultimately be able to reproduce life itself. Reproduction is an essential feature in any definition of life, and molecules that can replicate themselves are one step further along the way toward emulating life."

Craig Venter now plans to take existing genetic information and use it in new combinations. He doesn't hold back when describing his vision of where he expects this to lead: "It's going to totally transform the industry in the next 10 or 15 years. You'll be able to sit at your computer and decide what chemical reactions you want. The computer will then search for the right combination of genetic properties and write the chromosome you need. You'll insert it into a cell and off it goes. You'll be able to use bacteria for any purpose you can think of. It's going to totally alter what we do on this planet."

Venter's vision may come true one day, but creating genuinely new forms of life will take more than simply combining standard building blocks. Once the race to produce the first artificial living cell has been won, a great deal of research will be needed to discover how to create hitherto unknown forms of life. We will need to learn how the properties of living organisms arise from the complex interplay of regulatory mechanisms. We will only be able to create useful and viable life when we understand exactly how it works. Such knowledge would certainly deepen our insight in existing forms of diagnosis and therapy.

The disentanglement of DNA and the cellular control mechanisms has already deeply influenced medical therapy. Predictive genetic tests are now available for many conditions. For some of them, effective interventions are available to reduce the risk. This makes it possible to adjust medical therapy to personal characteristics of the patient. This is only the beginning, as prices for DNA sequencing are still dropping rapidly. Perhaps one day our understanding will be so complete that we can not only heal ailments, but also effectively re-create life. Craig Venter is likely to have retired by then, although you wouldn't bet against finding him out there on his yacht, *Sorcerer II,* for a good few years after that.

4.3

PREPARING FOR PANDEMICS

The first draft of this chapter was written before the pandemic alert for the 2009 flu was launched. Since then, terms such as *swine flu, Mexican flu,* or *H1N1* were constantly in the headlines. We witnessed the first really worldwide outbreak of a new influenza strain. Events went faster than we foresaw in our original text. We had started the chapter with an imaginary scenario of an outbreak in 2013 not in Mexico but in the East Java, Indonesia, city of Malang. It was not really meant as a prediction but merely a little story to show the consequences of an outbreak. We wanted to show how disruptive the outbreak of a new disease might be. We described all the things that we are now familiar with: doctors who aren't particularly worried in the beginning; people that live close to their animals and pick up viruses; patients in hospitals with high fever and severe cough; pharmaceutical companies anxious to peddle expensive vaccines.

Then we invented some struggle between the Indonesian authorities and the World Health Organization (WHO) about blood samples. That reflects the reluctance of developing nations to cooperate in the production of vaccines they can never afford.[1] In our story, the rest of the world ignored this imaginary outbreak and was oblivious to the rising death toll and the diplomatic wrangling. That's just like the start of the 2009 flu that probably haunted Mexican villages for many weeks unreported. In our story, the silence was broken when two nurses died in Perth, Australia. The media seized on the story immediately with yelling headlines. In the week that followed, dozens of new cases were reported in Indonesia, Australia, and Singapore, together with the first suspected case in New York.

Then there follows all the health humdrum that we are now so familiar with. The WHO has got hold of the flu virus and is preparing to produce a new vaccine. However, the epidemic spreads like an oil slick with the virus striking one major city after another. Antivirals change hands over the Internet for huge sums despite doctors' warnings that the drugs only work if administered within a few hours of infection. The WHO warned that it would take more than half a year to produce a vaccine that would halt the spread of the epidemic.

There was, however, one important difference between our story and the 2009 pandemic. Our virus was more deadly. We had a deadly strain of the H5N1 avian flu virus in mind—one that had became contagious in the human population. The invented patients in the Malang hospital all rapidly died of their fever and cough. The Mexican swine virus was relatively harmless. Previous outbreaks were far more deadly. A global outbreak of Spanish flu in 1918 cost between 50 and 100 million lives[2]—far outnumbering the 15 million who perished in World War I. The 1830 flu pandemic was no less lethal. Events in 2009 would have unfolded in quite a different way if the virus had been as deadly as the 1918 strain.

Our global infrastructure would have been disrupted. Imagine, for example, what truck drivers would do after a colleague died. Most would opt to stay home. As a result, the hospitals can no longer count on daily supplies of oxygen or medicines. Highly efficient, just-in-time logistics has led in recent years to very low stockpiles, leaving hospitals extremely vulnerable. The electricity supply would also begin to waver; power-plant managers desperately encourage technicians to turn up to work, but many of them will be too afraid to leave home. Meanwhile, some countries would seal their borders. That's an understandable reaction, but it leads to further disruptions. For example, 85 percent of U.S. pharmaceuticals are manufactured abroad.[3] There is enough oil to last for just 6 weeks.[4] The days are gone when a society could cut itself off from the outside world.

The breakdown of logistic systems would also affect the production of vaccines. The facilities are the same ones that produce the "normal" flu vaccine administered—mostly in the autumn—to some 350 million people worldwide every year.[5] The vaccine virus is cultivated in embryonated eggs, requiring some 350 million fertilized eggs.[6] Worldwide, around 6 billion eggs are produced annually, so there is a lot of raw material for vaccine production.[7] Data that were released a few months before the 2009 outbreak show that production can be stretched to 2.5 billion doses a year; thus, it would take 4 years to satisfy global demand of two doses a person. This capacity is expected to increase with a factor two to six until 2015, but even then, it would

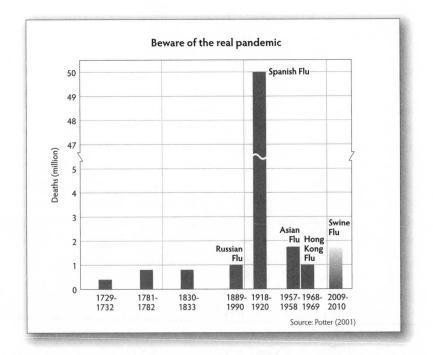

Beware of the real pandemic

We don't know when we'll be on the receiving end of a pandemic comparable to the Spanish flu. What we can say, however, is that the resultant disruption will be greater than it was in 1918 because of the far-reaching globalization of human society that has occurred since then. To prepare effectively for a real pandemic, we need to adopt a global approach and produce vaccines more efficiently. *Source*: Potter, C. W. (2001). A history of influenza. *Journal of Applied Microbiology, 91*, 572–579.

take 1 or 2 years to produce all the vaccines.[8] Ideally, that is. When the transport system is overstretched, we will need more time. And chickens may be susceptible to the same flu virus so that reliable supplies of eggs may be unavailable.

NOVEL VACCINES

The flu virus evolves rapidly, which is why it is able to strike again every winter. For the most part, the results are not unduly dramatic. Most people have antibodies left over from previous flu seasons which continue to offer some protection against the latest mutations of seasonal flu viruses. In 2009, however, an entirely new strain emerged as a virus made the leap from animals to humans. We have absolutely no resistance to a new strain like this, and the result is a pandemic. We can't say how likely any such event is to occur. All we know is that there were seven major flu epidemics in the past two centuries. Perhaps the next pandemic will strike next year or later. And maybe it won't start in Mexico but in China or Vietnam. There are many places in Asia where people and animals live in close proximity.

The frightening thing is that in conditions such as these, new flu viruses emerge. There is no way to influence the natural reservoirs where the viruses reproduce and mutate. Any new influenza pandemic could kill hundreds of millions of people, overturn global infrastructures, and bring the global economy to a standstill. It might conceivably put an end to modern civilization.

David Fedson is one of the people warning against a global influenza catastrophe. A former professor of medicine at the University of Virginia and retired medical director of Aventis Pasteur MSD in France, Fedson has dedicated his professional life to researching influenza vaccination. He now lives in a 350-year-old house in Sergy-Haut, a small French village near Geneva, Switzerland. His compact, friendly appearance belies his sobering message that a catastrophic flu pandemic might strike any moment. What preparations has he made in his mountain village to protect his family against any such pandemic? "I haven't," he admits. "If I had any antiviral medicine, it couldn't possibly be enough to distribute among all the people in the village. What would I do if my neighbour's child were dying? I couldn't live with the idea of keeping it all for myself."

Not all of his colleagues did the same. In a sample of flu experts during the 2009 epidemic, half of them admitted that they had taken precautions such as acquiring a supply of Tamiflu for their families. They said they worried that local hospitals could not cope in the case a nastier strain would emerge.[9]

"This isn't something you can prepare for on a personal level," David Fedson reacts. "It's our governments that ought to be prepared. But they aren't. In a severe pandemic, we would need billions of doses of vaccine, and we'd need to produce them quickly. But if a new pandemic were to break out tomorrow, we would probably be helpless." The 2009 swine flu pandemic shows this clearly, Fedson continues. "It has not been catastrophically severe, which is good. But the events show the inability of the international community to work together in the face of a common health threat. All the confident statements of health officials that we could respond effectively have not been borne out. Our ability to quickly detect the emergence of a pandemic virus was completely sabotaged by its emergence in Mexico. We have a hopelessly ineffective system of virus surveillance in pigs despite evidence from virologists going back to the late 1990s that a new pandemic virus could emerge in pigs. The world's vaccine companies and regulatory agencies have continued to use 40-year-old genetic reassortment techniques to make the strains for vaccine production. This is far more laborious than the use of reverse genetics. We have seen that the process of making vaccines is difficult to accelerate and scale up. Production takes at least 9 months. By the time you're finished, you'd only be able to vaccinate the survivors. We need a scientific breakthrough in order to speed things up. If the H1N1 virus would have mutated to become more virulent, we would have found ourselves in a terrible predicament."

"There are several steps we could take to prepare ourselves better. The least we should do is to consider strategies for producing more doses from smaller amounts of vaccine virus. We could add chemicals called *adjuvants* that would enhance the response of the vaccine. Research has indicated that this could reduce the required amount of virus in each dose to just one-tenth of the current amount." These techniques were available during the 2009 outbreak, Fedson remarks, but regulatory authorities in vaccine-producing countries didn't allow the use of them. "This refusal limits the numbers of doses that will become available, which means that many countries will not be able to get their supply of pandemic vaccines. You really have to decrease the amount of antigens in a vaccine to the lowest possible dose in order to protect as many people as possible."

There may also be ways of decreasing the dose even further, Fedson continues. "The resultant vaccine might be less effective from the point of view of the individual, but you'd be able to vaccinate more people, which might be able to achieve a higher protection rate for the population as a whole. That's a controversial idea, as it places population protection above individual health, and it's never been rigorously tested. But we need to consider it and to study it further. It might not give us enough doses for entire populations, but it would probably give us enough doses to vaccinate workers who are critical to maintaining the social infrastructure before it grinds to a halt. If they've been vaccinated, people can at least be confident that their societies won't collapse."

Adequate protection of a larger proportion of the population will, however, require the development of novel vaccines, Fedson believes. "Maybe we can use live attenuated viruses instead of the inactivated ones that we now use for flu vaccines. That's been shown in the laboratory to offer broad protection for mice and ferrets. The advantage is that you would need only one dose per person, it could be administered via nose drops, and you wouldn't have to use syringes. You could produce several billion doses in a few months in existing facilities for producing inactivated human or animal flu vaccines." There are other new ideas as well. "One possibility is to produce particles that resemble an influenza virus. Elegant bioengineering technology exists to do this, but it needs to be developed further."

These new approaches would radically disrupt the business of the big vaccine companies and their sponsors, Fedson notes. "The technology for egg-based production of inactivated flu vaccines stems from the 1950s and hasn't really changed much since then. Until recently, there's been little reason to do so. It's been perfectly adequate for producing the seasonal flu vaccines that we get each year. In the past decade, vaccine companies have invested billions in expanding vaccine production capacity with this classical technology. It will take several years to complete the projects that are now under way. Consequently, alternative production technologies for new types of vaccines have virtually no chance of being adopted, at least not in the near to mid-term." These new waves of investments don't only go into classical egg-based plants but also into newer production technologies that use cell culture. That makes vaccine production independent from eggs, and it is also easier to scale up. But it will not speed up the time required for production itself. So it's not much of an improvement."

Another factor is the high degree of specialization of the people involved in vaccine production. "It's a small elite group of scientists, policy makers, and company executives who are only familiar with vaccines and antivirals," Fedson confirms. "That makes it difficult for them to consider alternatives."

LET THE POOR LIVE

We're not only failing in terms of vaccine production; a massive program also needs to be established to distribute and administer those vaccines to different populations. It would be very difficult in developing nations, Fedson fears, because a mere nine countries produce almost all the influenza vaccines, and all of them are located in the developed world.[10] "Nonproducing countries have to import all their vaccines. But producing countries will vaccinate their own populations first. A pandemic vaccine is never going to make it out of the country of production. So people who live in countries that don't have their own vaccine manufacturers—and that's more than 85 percent of the world's population—will have little prospect of being vaccinated. Global vaccination requires such a degree of international organization that it is not even being considered." For that matter, it would be extremely difficult to supply the world with enough syringes to deal with pandemic vaccination. We would need billions of syringes. For the United States alone, you would need an extra 600 million to administer two doses to its 300 million inhabitants, which would take 2 years to produce.[11]

"Existing programs for pandemic vaccination and antiviral treatment have nothing to offer people in low- and middle-income countries. As a result, the rich will live, the poor will die, and the wounds to global society in the pandemic aftermath could fester for decades," Fedson states candidly. The silence in the media during the 2009 pandemic illustrates this point, he remarks. "News reports have generally paid little attention to countries that got no vaccine. They didn't ask basic questions about this unprivileged part of the world. Is this because these countries can't afford it? Because they weren't quick enough to order so that they are now at the end of the queue? Do these countries think the impact of swine flu will be not severe and disruptive? We haven't heard from these countries, and the silence reflects our overall lack of a detailed global perspective on what the disease is doing throughout the world. There may not have been millions of victims, but still, the death toll in those countries is probably considerably higher than in a normal flu epidemic. The fact that millions more are being infected means that the numbers of deaths went probably up. This brings up the question of how this will go when there is an outbreak with a really high fatality rate."

"We don't have the means to combat a pandemic globally. Commonly accepted ideas about how to confront the next pandemic mostly rely on existing technologies and centralized vaccine production. Yet no centralized complex technology can meet the challenges posed by a pandemic. Our current top-down approach is slow, complex, and difficult to organize and manage, and that is its fundamental flaw. It reflects a misunderstanding of what is needed. Instead, we need to identify technologies that we can share with developing nations. We need a bottom-up approach based on ordinary people and existing health-care systems, one that is based on abundant supplies of inexpensive generic medications that would be available worldwide on day one of a pandemic."[12]

Statins, fibrates, and glitazones are good candidates, Fedson believes.[13] They would work not by countering the virus but by shoring up the host response to the infection. "They're used preventively for patients at risk of heart attacks, congestive heart failure, and strokes and for treating patients with diabetes. Influenza is associated with these events, and clinical and laboratory studies indicate that these agents could reduce hospitalizations and mortality during a flu pandemic. Moreover, they are inexpensive and are already being produced in developing countries. They might not be the ultimate solution for an influenza pandemic, but I believe they offer the only possible strategy for limiting the damage on a truly worldwide scale."

LOOSEN THE NETWORKS

Decentralization is the only realistic response to a global breakdown of infrastructure and a surge in nationalistic selfishness. That's not only true for the production of pharmaceuticals; it's also key to the survival of many other parts of our tightly interwoven international networks. Worldwide interconnectedness has increased to the point where we are utterly dependent on our infrastructure. Networks are so tightly meshed that any disruption rapidly cascades through many sectors. Without electricity, no water can be pumped into the cities, food can't be kept or processed, the trains carrying coal can't run, the mines stop operating, refineries have to shut down, the Internet and other communication lines fall silent, and the global financial system seizes up. Each of these factors will then affect other sectors in turn. No diesel means no farming; no finance, no industry; no flights, no pharmaceuticals.

Reducing global shocks requires us to loosen these networks. It's interesting to see that the Internet is far less susceptible to sudden changes. The network itself is likely to keep functioning if computers were to come under attack of computer viruses because its control structure is decentralized. A computer pandemic might cause some network nodes to fail, but communication would then be automatically rerouted. Only a directed attack on some crucial nodes near long-distance connections could cause a digital traffic jam (see chapter 3.2). The robustness of the IT infrastructure means that new programs can be distributed quickly to repair the security flaw. They are developed by specialized firms that respond immediately and work extremely fast to provide a remedy. Companies recovering from the infection of their computers can quickly reconnect with the network and continue their business.

In preparing for the next flu pandemic, we should imitate the robustness that is a feature of other critical infrastructures. Preparing for major shocks means loosening our global networks, and decentralization would make our world more stable. People who use solar cells and wood-fired ovens would be at an advantage if a pandemic were to break out because they are only partially dependent on the electricity network. Cities with strategic food reserves will be better able to survive. Small production units, without global supply chains, will be able to continue longer and recover faster.

A certain amount of centralization will remain necessary, David Fedson thinks. "Production and distribution of generic agents like statins will be needed. Even though they are produced in countries like India, you will still need to distribute them throughout the country and to other countries nearby. But in many regions, there will already be an abundant supply on day one of a pandemic. After all, these agents are commonly used to treat ongoing conditions on a daily basis. These amounts are probably enough to treat 2 or 10 percent of the population—those who are critically ill."

4.4

QUALITY OF LIFE

How old will our children live to be? 120? 150? The average human life span continues to lengthen, and more and more of us will enjoy a long life. A substantial proportion of today's children will one day celebrate their 100th birthday, whereas back in 1900, half of all human beings were dead by the age of 37. Life expectancy in the Western world has advanced with remarkable speed, which means the number of old people is also increasing rapidly. A century ago, a mere 1 percent of the world's population was aged older than 65. By 2050, that figure will be about 20 percent. Babies born in 2010 will live an average of 20 years longer than those born in 1950.[1] Life expectancy increases 3 years for each decade that passes, reflecting ongoing progress in technology. The necessities of life are provided more efficiently than they were a century ago, certainly in the West. There is enough to eat, and we are well clothed and sheltered. Advances in medical science mean we can live longer without falling victim to disease. And if we do get sick, we can survive longer. Chronic illness, heart conditions, and cancer are no longer necessarily a death sentence.

People in the developed world now live so long that the main causes of death for those under 50—violence and suicide—lie beyond the reach of medical technology. We only become dependent on medical intervention later in our lives, as the age at which we begin to "break down" has risen progressively over the past century. Today's old people are much sprightlier than their counterparts in the past. A person now aged 75 frequently has a similar level of health, vitality, and *joie de vivre* as a 65-year-old two generations ago. We wear out less, our living conditions are better, and prompt action is taken if something goes wrong. Most important, many people believe it is worthwhile to live longer as we can enjoy the extra years in good health and pleasant circumstances. As we discuss in this chapter, however, we face the possibility that staying alive longer could also entail a set of completely new problems.

How long can our life span be extended? For the vast majority of the period in which *Homo sapiens* has existed, humans have barely survived past the age of 40. Throughout the last century, however, life expectancy was steadily raised, and there is no sign yet of a possible upper limit. Human beings are tougher than we thought; our bodies don't appear to have a built-in sell-by date, whatever scientists may have long suspected. We're not fitted with some kind of time bomb primed to go off when we hit the age of 80, 100, or 150. Self-destruction at an advanced age would not serve any obvious purpose in terms of the survival of our species.[2]

There is a fixed relationship between body size and life span. Elephants live for 70 years, cows for 30, and mice for just 2 years. Large

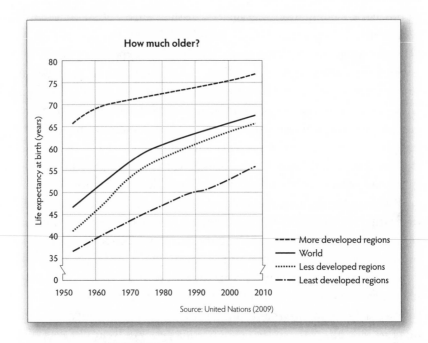

The proportion of elderly people is increasing everywhere and not only in the most developed regions. There's nothing to suggest that future generations won't enjoy an ever-lengthening life span. This is first and foremost a triumph of medical technology. Yet it also has immense social implications because the resources simply aren't there to provide every old person with adequate care. Technology may provide an important part of the solution.

animals also have slower metabolisms than small ones, as we will also note in chapter 5.4. Because everything operates more slowly in larger animals, they wear out less rapidly, too. Differences in the immune system are another factor. DNA defects are repaired more effectively in elephants than they are in mice. Larger animals have more complex immune systems. Their bodies have the necessary room to store millions of different defensive cells in their blood. Smaller animals don't have that luxury and so have to combat diseases in a less refined way, which reduces their likelihood of survival.[3]

No matter how refined the immune system, however, a certain amount of damage will always go unchecked, and this accumulates in the course of our lives. No single instance is critical in itself, but in combination, it can lead to the steady degradation and disruption of the organism. Damaged elements interact and stresses increase, eventually leading to an overall breakdown. Our bodies ultimately fail as complex systems. In the course of our history, we have managed to postpone this breakdown for longer and longer. In that respect, human beings are a striking exception to the rest of the animal kingdom. Based on our body weight alone, our species should have a life expectancy of about 15 years—roughly the same as a pig or sheep of the same size. Yet *Homo sapiens* has similar survival chances to those of an elephant, which weighs 100 times more than us. We can overcome our bodies' malfunctions until we reach an advanced age, at which point they suddenly trigger a fatal crisis.

We talked about these issues with Steven Lamberts, a Dutch physician who specializes in aging and former rector of Erasmus University in Rotterdam. Breakdown often occurs abruptly, as is so often the case in complex interacting systems. "We used to have a longer period of gradual decline before finding ourselves in genuine need of assistance," Lamberts notes. "But the transition from healthy to ailing has become sharper. Things suddenly start going wrong around the time we reach 75. People become dependent. And from that point on, they often require many years of being cared for intensively. That worries me."

Perhaps we could reduce the deterioration of our bodies by implanting new tissue to counter the influence of damaged DNA. Or maybe we could further extend our life span using therapies capable of beefing up our immune systems. Our internal regulation would then break down a little later. Lamberts believes, however, that such developments would simply make what comes next even sadder. Advanced age isn't synonymous with a happy life, he says. As a physician, there are moments when wards full of elderly people move him to tears. "I see no point in trying to raise our maximum age. All we're doing then is to make the final, dependent stage of our lives even longer. It makes much more sense to try to avoid getting into that situation or, if it does happen, to make it as bearable as possible. It's an issue that's crying out for attention but one that we totally ignore. In my position, I'm profoundly disappointed to see old people confined to hospital wards. People who have worked hard all their lives lie there, five or six to a room, waiting to die. I don't want to see them in bed. It's unacceptable that tens of thousands of people end up lying in their own urine and feces. But it's not realistic to think we can find the personnel to take proper care of all those people around the clock."

That's also an issue in the homes of the elderly, Lamberts thinks. "In the next 20 years, the number of genuinely elderly people will double in many Western countries. It's hard to imagine that we'll have enough home-helps to lift them all out of bed twice a day. We don't have that many pairs of hands, which means we'll have to look to technology. Quality of life has everything to do with independence and the chance to lead a social life. So it's a question of technology to help us see, hear, go to the toilet, comprehend, communicate, and get up. I can visualize a series of aids, which are urgently needed and which could be readily manufactured, that wouldn't be too expensive and would considerably increase the quality of elderly people's lives."

Lamberts believes these things don't yet exist because technological development isn't geared toward the elderly. "Pictures of the high-tech homes of the future invariably show healthy young people who don't need any of that stuff. They're perfectly capable of finding the light switch. Why would they want special sensors to do it for them? We have to design comfortable homes for our old people. It's technology that would actually save us a lot of money because it would help the elderly remain independent longer. If a group of engineers were to focus on it seriously, we could achieve a breakthrough. Our society urgently needs technological assistance for old people, yet it never seems to happen." In Steven Lamberts's view, physicians don't pay enough attention to the issue either; as far as they are concerned, their job if someone contracts pneumonia, for example, is simply to prescribe antibiotics. Yet much more has to be done before an old person is capable of living independently again after such an illness. "It's not a medical issue; it's a question of how we support our elderly population. If our society manages to improve the quality of the final stage of life, we will have achieved something very special." Lamberts had his wish list ready when we talked to him. He proposes specific technology to help us as we get old and has a series of technical suggestions to address the most pressing problems.

SEEING AND MOVING

Steven Lamberts thinks, for instance, that technology can help combat blindness. One of the main causes of deterioration in old people's vision is macular degeneration. One-third of 75- to 85-year-olds suffer from the condition, which results from the gradual accumulation of fat behind the retina, damaging the light-sensitive cells, and eventually leading to blindness. Yet only the central part of the retina is affected—the area we normally use to read. The edges remain intact. Patients can often continue to read using a special prism that exploits the retina's remaining peripheral visual capacity. But this doesn't help them walk or boil a kettle. Heavy, head-mounted devices exist, with cameras and screens so that people can see things around them. "But there's no reason in theory you couldn't use microelectronics similar to those in your mobile phone. That would enable patients like this to go on taking care of themselves. Aids of that kind would contribute significantly to maintaining their independence."

New technology for the joints also features on Lamberts's wish list. Replacing a joint is not only expensive, but its impact is also temporary. Artificial hips and knees don't last for more than 12 or 15 years due primarily to the way they are attached to the bones. The artificial joint is fixed in place using metal pins. Over time, the bone grows thinner, and the pins loosen. What's more, artificial joints have only a limited range of movement; they make you walk a little stiffly. It also means that the joint lacks the flexibility to cope with unusual movements, causing it to wear out faster. This extra wear, combined with the loosening of the anchor points, means that the joint eventually has to be replaced. Patients fitted with an artificial hip when they are 75 will need a new one when they reach 90. The more old people there are, the more this will happen, so we need to extend the life span of artificial joints. There are two ways to do this. A great deal could be achieved, for instance, by using better materials, more sophisticated joints, smarter mechanics, and alternative fixing techniques. An even better solution would be to select natural materials to give the person's original joints a new covering—pieces of cartilage, for example, to reline a knee or hip joint. Tissue engineering should make that possible.

Our muscles also begin to fail as we get older, robbing us of the strength to get out of bed or go to the toilet on our own. Lamberts thinks a great deal can be done using mechanical aids. Metal hands are needed around the bed—lifts to help you raise yourself or get off the toilet. This will ultimately mean domestic robotization, as we saw in chapter 3.7. Precision tasks, such as putting on support stockings, are still very difficult for robot arms; leg sizes and patient behavior are too varied. But a bed lift with simple buttons could easily be produced using existing technology. This kind of assistance at crucial moments will keep people independent. Another technique is to stimulate the muscles. Paraplegics can take steps if strategic stimuli are delivered in this way. This is worthwhile in the case of young patients, as it saves on the care they would otherwise need for the rest of their lives. Elderly people often don't receive this kind of assistance, even though it would help them move step by step or get off the toilet. Steven Lamberts believes we need to come up with smart devices that will make the delivery of this kind of stimuli cheaper and hence viable for old people as well.

CONTACT

Deafness, too, sometimes boils down to cash. Congenitally deaf children receive cochlear implants—prosthetic organs that take over the function of the malleus, incus, and stapes, the elements of the ear that detect vibrations. It costs $80,000 to restore a child's hearing. Adults, however, often don't get cochlear implants, let alone old people with only a few more years to live. "There shouldn't be any deaf-mute people in our society anymore," Steven Lamberts thinks. "And we can also treat many forms of deafness in old age. The technology is expensive, but I find it unacceptable that a relatively large group of people is not being treated for inner-ear deafness. We need to find cheaper techniques so that more people can benefit from this kind of treatment." Technology can help to fend off loneliness, too, Lamberts says. "That's crucial to old people's sense of well-being. There are lots of communication devices for the elderly, but these often consist of little more than a panic button to alert the emergency services and at best a phone link to a support center. Great if you suffer a fall, but not much help if you simply need to chat. Young people phone, text, and message each other with or without Web cam pictures and broadband Internet connections. Old people have to make do with old-fashioned phone lines to talk to their children and friends. Visual information can strengthen contacts, especially when your hearing isn't what it used to be. A simple videophone for old people—that surely couldn't be too difficult to achieve."

The biggest challenge is to support elderly people as their cognition deteriorates. We know that mental exercises offer effective protection against memory loss; communication and social activities also slow down the decline in cognitive performance. Technology can help us, too, as our powers of memory begin to fade. "Visual and audible alarms, for instance, can remind us that it's time for a meal or to take our medicine. We need cookers that turn themselves off and kettles that make it impossible to scald ourselves. Cognition often declines very gradually before dementia sets in, and the consequences can, to some extent, be eased by technology. That's important, although it offers little real comfort. Technology can do painfully little to aid cognition as such," Lamberts remarks.

These are just a few examples of the breakthroughs—both large and small—that will be needed to help compensate as our faculties decline with age. There are other things that could make life easier. It isn't primarily a question of microbiology or life-prolonging pharmaceuticals as much as adding to humanity's existing toolbox. It's our brains, after all, that have given us a longer life span than that enjoyed by sheep or pigs. We can invent all kinds of tools to extend the human body, effectively making us larger and stronger than our size would suggest. In many cases, aids for the elderly are already technologically possible but will require cheap mass production for them to achieve a genuine breakthrough. And that will mean a different attitude on the part of designers. Rather than complex technical aids loaded with functions that drive up the price, they should be stripped to the essentials: simple hardware that can be produced in large numbers. It's a challenge our industrial designers should take up. It is also an example of a complex problem, which means different scientific and engineering disciplines will have to combine to achieve the necessary breakthroughs. Practical solutions will require the very best of our electronics, mechanical engineering, telecommunications, and control systems, as well as our medical science.

Part 5
COMMUNITIES

5.0

ENGINEERING SOCIETY

A storm blew up in Berlin in 1989 not far from the spot where much of this book was written. It all began in a small way with people attending weekly services at the local church to pray for peace. When the communist East German regime used violence to break up a demonstration, the church became a refuge for hundreds, and later thousands, of people. The society in question had grown rigid. To express it in the language of complexity, the social network became so tautly stretched that any shock was readily propagated throughout the system. The police repeatedly beat up the churchgoers, but the multitude failed to respond in the expected way. Instead of kicking and punching, they prayed and sang. They didn't display the anticipated logic of action and reaction, eventually causing the police to withdraw in confusion. The demonstrators created positive feedback, and as a result, the mass of people grew even bigger. "We were prepared for everything but not for candles," a police commander later commented.

The protests also confused the GDR's inflexible leaders. At the peak of the protests, an East German minister declared that citizens would be permitted to travel to the West. The chaos that ensued was so great that historians are still trying to unravel the precise sequence of events. On the brink of a critical transition, old forces dissipate and unpredictable movements can occur. This is a typical example of a small movement that can lead to much greater things, as we have also seen in other complex systems. Tens of thousands of people laid siege to the Wall. Exactly who eventually decided to raise the barriers has been lost in the fog of history. It was most likely a low-ranking officer at a border crossing who was no longer able to cope with the mass of people. To ease the pressure, he allowed a few citizens through the barrier. The effect was to throw gasoline onto the fire or, to put it another way, to create positive feedback that tipped the situation into transition. Within minutes, the crowd could no longer be restrained.

The only option left was to open the border once and for all. The Berlin Wall had fallen.

The peaceful revolution in the GDR can be understood intuitively in terms of a complex system that has moved far out of equilibrium. There is a growing group of sociologists who use the methods of complexity science to achieve better insight into processes of this kind. Taking the behavior of individuals as their basis, they attempt to model how this might give rise to collective phenomena. Their studies mostly involve societal movements that are more gradual than revolutions. They study, for example, how opinions and rumors spread, how clusters form, and how applause ripples through an audience. These processes then help us understand more complex situations, such as bouts of financial market panic, urbanization, and indeed, popular uprisings.

This emerging science of social dynamics is benefiting from the enormous amount of new data on human behavior that has become available in the last 10 years. Neural science is laying the foundations for a new understanding of human conduct and interaction on an individual level. At the group level, meanwhile, the new social scientists can draw on the mass of data that our digital society produces as traces of almost every aspect of human activity. The urbanization expert who contributes to chapter 5.3 cites the availability of all the mobile phone data from a major city, which allows him to monitor precisely how people form new personal networks when they resettle there. The resultant picture is far more complete and accurate than anything sociologists could ever achieve based on their traditional methods of questionnaires and manual counting.

Another new quantitative instrument is the ability to create virtual societies. These resemble computer games like SimCity in which individual citizens interact according to prescribed patterns. The resultant model society consists of an ever-changing network of interactions in which a large number of people respond to one another. The society unfolds within the computer and reveals the kind of collective behavior that might develop. It can be explored, for instance, under what circumstances an uprising will result in a revolution. Key episodes from the Russian Revolution have been simulated in this way.

All this offers a new quantitative approach to sociology. Old-school sociologists also used calculations, but lack of data and computing power obliged them to limit themselves to static situations close to equilibrium. In this new approach, the dynamics of a situation can be simulated taking account of nonlinear, nonequilibrium effects, too. This has shown itself to be essential in terms of understanding herd behavior and other "irrational" phenomena, which in turn makes it possible to study how markets crash or societies change. The new approach is already yielding an explosion of fresh results.[1] Many new insights have been obtained by outsiders—often physicists or mathematicians—who are trained to model complexity and identify patterns in large data sets.

We also encounter specialists in the following chapters who operate on the fringes of their field. One of their first successes was to produce accurate statistics for stock exchange crashes. Generations of economists have expressed puzzlement at these "statistical outliers." Now, however, major crashes can be understood in the same context as smaller stock market fluctuations. Other similar studies show how viruses spread, how people move through a city, and how the airline network evolves. Various secret services are no doubt performing similar research, too. It does, indeed, have the potential to help the leaders of both Koreas, Middle Eastern states, Myanmar, and anywhere else people have built walls to separate one group from another. Who do you need to arrest to keep a protest from getting out of hand? How do you break up a determined group of chanting people prepared to defy the police? How do you persuade people in the midst of chaos to follow their leaders once again? It's all very relevant information for authorities and protestors alike.

We begin this part of the book with two chapters focusing on the individual (chapters 5.1 and 5.2). In these chapters, we seek to answer questions about how our social environment affects us, how we learn, and how our brains are altered by hectic outside stimuli. We then step deeper into the agglomeration with some thoughts on the rise and fall of our great cities (chapter 5.3). Next, we describe how to cope with herd behavior in disasters (chapter 5.4). We round off with two chapters on human networks that span the globe. Economic networks are now so taut that the effects of a financial crisis are felt everywhere (chapter 5.5). That same interconnectedness, however, also means that countries are less likely to resort to military conflict against one another (chapter 5.6).

5.1

ESSENTIAL EDUCATION

The helplessness of newborn babies is very endearing. They can just about breathe unaided, but they are otherwise entirely unadapted and dependent. Babies can barely see, let alone walk or talk. Few animals come into the world so unprepared, and no other species is as dependent on learning as human beings are. Elephant calves, for instance, can stand up by themselves within a few minutes of being born. Most animals are similarly "preprogrammed." Female elephants carry their young for no fewer than 22 months, whereas we humans have to go on investing in our offspring long after they are born. Children need years of adult protection. They guzzle fuel, too; their brains consume fully 60 percent of the newborn's total energy intake. In the first year of life, the infant's head buzzes with activity as neurons grow in size and complexity and form their innumerable interconnections. The way the brain develops is the subject of the next chapter (chapter 5.2). Here we concentrate on the way we are educated from the first day on.

There is virtually no difference between Inuits and Australian aborigines in terms of their ability—at opposite ends of the earth and in climates that are utterly different—to bear children successfully. Other animal species are far more closely interrelated with their environment. Other primates have evolved to occupy a limited biotope determined by food and climate. Humans are much more universal. Every human child has an equal chance of survival wherever they are born. As a species, we delay our maturation and adaptation until after birth, which makes the inequality of subsequent human development all the more acute. Someone who is born in Mali or Burkina Faso is unlikely ever to learn to read.[1] A person whose father lives in Oxford, by contrast, might have spoken his or her first words of Latin at an early age. Inuit and aboriginal babies may be born equally, but their chances begin to diverge the moment they start learning how to

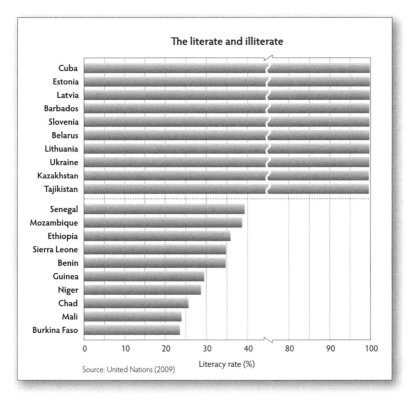

Education is one of the keys to any effort to solve global problems. The chart shows that countries with strong centralized governments have high literacy rates. But law and order aren't the only factors determining a population's educational level. Research in neuroscience, sociology, and computer science is also shedding fresh light on learning, including education in remote areas. Strong communal bonds and new communication technology are clearly very important, too. *Source: United Nations Development Programme Report 2009.*

live. We are not shaped by our inborn nature but by the culture that is impressed upon us by the people with whom we grow up. Learning makes children different, which is why it is unacceptable that over 100 million children of school age do not attend lessons. As one of its "Millennium Goals," the United Nations has committed itself to reducing that number to zero by 2015. Sadly, although a certain amount of progress has been achieved, it is likely that two-thirds of the eighty-six countries that currently lack basic education will fail to achieve it by 2015.

Education is often the victim of politics. It is odd that world leaders should be challenged about their use of torture, curtailment of liberties, and lack of fair trials but not about their education policies and the effect these have on mothers and their children. Basic health care is part of the same issue; mothers and children who are sick are hardly in a position to develop educationally. Schooling also suffers from a lack of teachers, educational resources, and proper understanding of juvenile development. A real breakthrough is needed to guarantee education and universal access to the fruits of human culture. New ideas from sociology, neuroscience, and computer science are changing our understanding of human learning.[2] These ideas could have a profound impact on education and hence on our culture in general. These insights might ultimately help us hold the human race together, ensuring equal opportunities for babies who are born equal.

IMITATION

Learning is an activity between people. Children use social clues to understand their world. Imitation of other humans is one of the cornerstones of child development. It saves a lot of trial and error because what someone else is doing is probably worth trying yourself. We're trained from an early age to recognize and respond to patterns. Babies mimic their parent's speech even before they have learned to talk. The babble of a Chinese baby is audibly different from that of an American infant. And if they were to switch cots, their language skills would never entirely catch up. Babies quickly begin to imitate their mothers' expressions and to stick out their tongues when their fathers do. A small child seeing its mother use a keyboard might also begin to poke at the keys. Behavior like this isn't instilled deliberately and might even be discouraged. And there's certainly no innate tendency to interact with a plastic object in such a way. The child simply imitates what it sees.

Imitation is built in. The regions of our brain in which perceptions are processed overlap the areas responsible for initiating actions. Experiments with children show that social interactions reinforce learning in countless different ways. Empathy, shared attention, and one-on-one coaching all help. Brains assume a more settled form in the course of a human life in response to thousands of examples. We store up procedures that enable us to function extremely effectively when a lot of things come at us at once. We can rapidly assess situations and make instant decisions without having to rehearse all the criteria every time. We jump out of the way, for instance, when we see a car hurtling toward us.

This is how our culture comes to be etched into a child's mind. If the mother can read and write, the child wants to do the same. The more developed a mother, the greater her children's opportunity for development. Culture gives children the opportunity to enter our civilization at a high level. A remote control and television are just as natural to a child as a tree is. Anyone who didn't grow up with the Internet will never entirely catch up. Every generation starts at a higher level with more complex knowledge. It's the way we learn that makes cultural progress possible.

THE COMPUTER AS A MODEL

Learning from experience is not something computers do well. Before they can get down to any useful work, they have to be spoon-fed programs that fix rules for every analysis and decision they are to perform. Computer scientists are trying to change that, looking for alternatives that mimic the way children learn. They have developed computers that you don't have to program in advance. Like newborn children, these machines begin with a clean slate, which is then filled in with observations that go on in turn to govern their actions. As we saw in chapter 3.5, these so-called neural networks can already be deployed in all sorts of situations where it is difficult to specify rules. Like people, computers can now make decisions based on vague criteria in areas such as controlling the temperature of a building or speech recognition. Thus, dealing with incomplete or half-understood information is no longer the exclusive preserve of living brains. Neural networks are not only useful for tackling problems that aren't readily programmable, but they can also help us understand how children come to grips over time with their complex environment. A computer is a laboratory for trying out how different learning rules work in practice. We can test, for instance, the most effective way for computers to handle complex input. They appear to learn more efficiently when they commence the learning process with just a few signals after which the complexity is gradually increased.

Computer learning can also be improved by incorporating an element of social interaction. In one experiment, a computerized doll was programmed to seek correlations between its own behavior and changes in its environment. Within minutes, it noticed that its own cries were invariably followed by sounds from the pale ellipsoid positioned in front of it. The doll had learned that faces have a certain significance. The same human face can then be used as a guide for attempting to classify more complex interactions.

Experiments like this are bringing a new dimension to the science of learning, as are the new insights emerging from brain research. Neuroscientists can see with increasing accuracy how information from the external world is conveyed to the brain and stored there before reemerging. This creates an opportunity to test education theories. Teachers have traditionally had to work on the basis of intuition, beliefs, and subjective observations. We now have the chance to explore experimentally the sometimes contradictory ideas of prominent educational theorists like Maria Montessori, Rudolf Steiner (Waldorf education), and Helen Parkhurst (Dalton School), thereby creating a new scientific foundation for pedagogy. Although this fresh approach is still in its infancy, breakthroughs in this area could deliver new insights that will make it easier to learn. That could in turn provide children—including youngsters in isolated regions—with more opportunities for early socialization in our complex world.

LEARNING COMPLEXITY

Hans van Ginkel has promoted learning in every corner of the world. The Indonesian-born Dutch professor originally taught human geography and planning. In 1997, however, he was appointed rector of the United Nations University in Tokyo, at which point his focus switched to learning in the developing world. Now retired, this former UN undersecretary and president of the International Association of Universities remains active in a wide range of pedagogic missions. "Education ought to address our world's complexity more effectively," he says. "Both the number of interactions and the distance over which communication occurs are increasing. We are aware of and see more people; we realize that the world has become more complex. A farmer in Ghana isn't just involved with his local market. He knows something about EU subsidies and has to take account of food production in the United States. Everything is linked to everything else; that's the essence of globalization. A world order is emerging that is characterized by connectivity, change, and convergence. Our children will have to learn, for better or for worse, to live amid this growing complexity. Viewed from a simple world, the steps toward complexity are very big. The human mind can only make limited jumps. You need to incorporate complexity in education from the earliest age; otherwise, it won't succeed. It's an important task that education has to fulfill. People need to learn how to think about complexity and scale. That runs counter to the ideas of certain educationalists—and a lot of other people who set the tone in our society. Today's education and media often place particular stress on simplicity. What we really need, however, is a better understanding of the *complexity* of technology and society."

"Complexity isn't only a question of different disciplines but of different scales, too," van Ginkel explains. "Education should help people develop a sense of scale. You can only learn how things interrelate as everything gets bigger if subjects are presented in an integrated way. Learning about the interrelationship of things should be at the core of what the school does. Part of that is learning to expect the unexpected. Complex systems are always different. Natural disasters aren't caused by a single event but by a combination of things. Early warning systems have limited value when they are based only on the experiences of earlier events. It's the same with many things in our society: Linear knowledge from preprogrammed curricula doesn't help in such instances. You need perspective and responsiveness."

That's not something you can learn through the straightforward transfer of knowledge, van Ginkel argues. "And you don't learn it by looking things up in Wikipedia either. Connecting villages to the Internet really isn't the best way to improve education. It's often claimed that through ubiquitous computing, the information society gives people the opportunity to make properly founded choices. If we can't place that information in context, however, it simply unnerves us. There's no shortage of information—far from it. What's important is knowing how to use it. That's one of the reasons online education has been doing badly in Africa. Initiatives like a virtual university don't work because they're alien to that society. Even people with Internet access can't really benefit from that type of education. Lessons broadcast on the radio do much better; they're simple to produce locally, they have a wide broadcasting range, and you can listen to them in groups. That shows that you have to use local institutions to introduce new ideas. Things that come from outside are less effective than things that come from within the group itself. Society has to be able to grow with them."

The new science of learning can help in this instance to make the right choices. "Virtual education of this kind places people outside their learning community. What's more, the budget for existing education structures is declining because of parallel initiatives like this," Hans van Ginkel warns. He is in favor of expanding that learning community. "As the world becomes more and more connected, you have to ask yourself whether it will be able to flourish if the gap between North and South persists or grows even wider. Sharing money isn't enough to equalize education levels. We have to share our thinking—our brains. Education is a profoundly human process. It takes more effort to share our thoughts than it does to donate money. Although volunteers from richer countries do their bit, they're often very inexperienced. Still, there are positive trends, too, with retired people and employees on sabbatical sharing their knowledge and experience. Twinning with relevant institutions abroad and targeted joint programs are also important."

Hans van Ginkel thinks, moreover, that the involvement of these experienced volunteers and institutions is a clear demonstration that the world doesn't revolve purely around money. "It's good to show people that positive things aren't always expressed in terms of cash. You have to have ideals. We don't want people whose ambition is simply to design better television screens; what the world needs is individuals who want to put their ideals into practice. The greatest threat to society is when people's ideals are taken away from them. You have to know that the world will open up to you when you finish your studies. That's the real global challenge."

5.2

MAINTAINING IDENTITY

Baroness Susan Greenfield's origins are humbler than her title might suggest. Her father was a machine operator in an industrial neighborhood of London. In Britain, unlike many other countries, it is possible to earn a peerage through your own merits rather than pure heredity. Lady Greenfield is a leading world authority on the human brain. She is concerned that technology has invaded our lives so profoundly that it has begun to affect the way our brains operate and hence our very personalities. "People are longing for experiences rather than searching for meaning," she says. "They live more in the moment and have less of a sense of the narrative of their lives—of continuity. They lack a sense of having a beginning, a middle, and an end. They have less of a feeling that they are developing an identity throughout their life with a continuing story line from childhood, youth, parenthood, to grandparenthood. The emphasis is more on process than content. You now have people who are much more 'sensitive' rather than 'cognitive.'"

Susan Greenfield identifies one of the causes of this development as the impressions our brains receive from a very early age. Modern life, she argues, with its hectic rhythm of visual impressions is very different from the past, in which she includes her own childhood in the 1950s and 1960s. It's in our youth that our brains are shaped: They grow like mad during the first 2 years of life, developing a maze of connections. And in the years that follow, they remain extremely nimble, forming new connections rapidly and changing in response to our surroundings. It is very much the world around us during infancy, childhood, and early adolescence that determines the outcome of this stage of brain formation. The brain displays an immense degree of what Greenfield likes to call "plasticity" during this stage; connections are formed as and when they are needed.

The foundations of Baroness Greenfield's own personality were laid in a similar way during her youth. The way young Susan devoured books shaped her brain at that stage, strengthening her imagination and her ability to cope mentally with extended narratives. Perhaps her future professional interest was formed then, too. The rabbit's brain she once held in her hands, for instance, is bound to have made an indelible impression. A fortunate convergence of nature and nurture made her a brilliant scientist, who rose to become professor of synaptic pharmacology at Lincoln College, Oxford, and former director of the Royal Institution. As a life peer, Lady Greenfield is also a member of the House of Lords, and she has published several popular books on brain research.[1]

WE ARE COMPLEX

The network of 100 billion neurons that forms our brain is one of the most complex structures we encounter in this book. The average human adult has 500 trillion connections in his or her head (5×10^{14}). Susan Greenfield likes to compare this with other complex networks, the similarities with which are, she says, very striking. Take the network of relationships we have around us. Like neuronal connections, they strengthen through use and with more intense input. They are constantly changing, especially during our youth. It is more than a superficial metaphor, she argues: Both kinds of network influence one another. The relationship is clearly apparent when the network underlying our brains begins to break down with the onset, say, of Alzheimer's disease—Greenfield's specialty as a neuroscientist. Brain development is actually shifted into reverse when dementia strikes. Connections in the brain are lost, and we progressively lose the skills we acquired in the course of our lifetime. Our personalities fade along with our ability to connect specific events into a logical narrative. Eventually, we see only generic faces around us that are no longer connected to memories. As our brain network breaks down, our social networks disappear, too.

A similar breakdown can occur under the influence of drugs that alter the fine balance of chemical interactions through which the brain communicates. In some cases, they merely block communication within the brain; more often, however, an excess of hormones like dopamine overstimulates neuron activity, thereby impairing a controlled brain function. We lose the ability to reason about the consequences of our behavior and start to live solely in the "here and now." Something similar happens during sex, when dancing, or during a roller-coaster ride. "It makes you feel good," Greenfield confirms. "But too much of it may damage your ability to reason beyond the thrill of the moment. You then lose connections in the brain."

It is this impairment of our brains that is Greenfield's greatest concern. Young people's thrill-seeking behavior can produce a dopamine overdose. She cites the example of computer games. "It may be a game where you have to set free a princess who is locked up. But in the computer games that you see now, it isn't about that person—you don't get to know her. It's all about the thrills you get in the process." And those thrills release dopamine into the brain, making you want to continue. "If you are surrounded by fast-paced multimedia, with very strong emphasis on the senses rather than on the cognitive element, then the brain will process it accordingly. It will increase your ability to have fast reactions, but you won't have sufficient connections supporting the meaning or the ability to relate one thing to something else. That's sad because it's a diminution of identity—of the sense of who you are. You can't define your identity in terms of the sensation of the moment. Identity is about meaning and significance."

Susan Greenfield believes that the amount of time people devote to such games makes it even worse. "Never before have so many human beings had so much free time, apart from maybe in monasteries. Until recently, daily life for most people was full of drudgery. You didn't have time to think too much, and you also had a shorter life span. Now is the first time in history many humans have the comfort and freedom from pain that allow us to devote our time to other things than our immediate survival. It's worrying to see that it's also the first time that grown-up people have begun to play games on their own as a child would. They don't use games as a vehicle for socializing as generations before them did. If all we have achieved as human civilization with all our science is sitting behind a screen to have some private thrills, I find that very sad. It is infantilizing adulthood."

THE NOBODY SCENARIO

These observations have led Baroness Greenfield to categorize human behavior into different personality archetypes. A person whose identity is formed by his or her relationships with others is a *Somebody* in Greenfield's classification. The brain of a Somebody has many connections, but they are constantly being rewired through impressions from the outside. The individual is how others see that person. The *Anybody* personality is a more conservative character. This person also has a well-connected brain but a less changeable one. The life of this type is more ritualized; its identity is less conditional and more uniform. It is more likely to be led by doctrines and might also be thought of as a fundamentalist. Western history from the Enlightenment to the end of the twentieth century has basically been a struggle between the Somebody and Anybody personality types—between individualism and collectivism; between the adaptivity demanded by the free market and the certitude offered by totalitarianism; between the overwhelming choices of democracy and the clarity of dictatorship.

The twenty-first century has produced a third archetype, Greenfield believes. It is a personality type with less extensive wiring in the brain, and one that lacks a framework capable of giving meaning to the person's surroundings. This *Nobody* archetype merely has experiences rather than overarching stories. "If there is no one to tell you the greater story, you won't learn to think metaphorically. Education is about relating something to something else. If that isn't shown to you, you won't be able to invent the greater story yourself." The Nobody mentality is reinforced by an excess of dopamine—the kind of kicks we get from computer games and a rapid succession of visual incentives. "That's both good and bad," Greenfield thinks. "The new generation has very agile brains in terms of multitasking, motor coordination, and seeing patterns. But they lack the deep meaning to what they are doing. As people live more in the here and now, they take more risks and have less sense of identity. There is more emphasis on the literal and taking things at face value rather than thinking metaphorically and symbolically."

Greenfield thinks there is already evidence that relationships are changing. "Someone told me she had 900 friends on Facebook. That's impossible: You can't have 900 friends. It devalues the very concept. A friend is someone with whom you take a long walk when you just finished a love affair or when you lost your job. It's not someone with whom you have just one line on Facebook." Taken to its extremes, she argues, people will no longer define themselves and others as having a separate identity. The whole story of what makes a life unique will disappear, effectively creating a society in a state of dementia.

Greenfield doesn't believe that any of these three archetypical personalities can give us the fulfillment and sense of self we need in the twenty-first century. But she also sees a fourth type becoming more important, the *Eureka* personality—a person whose brain is sufficiently plastic to abandon old connections and to form new, hitherto unimagined links. It is a creative personality that is not molded by outside impressions but that reconnects its brain through its own effort. Talented scientists, artists, and composers belong to this category because they use knowledge and understanding of the knowledge to generate new ideas. Knowledge has limited value if we are not able to place it in a broader context. Understanding the issues behind knowledge is subsequently the platform for creativity. The Eureka types are thus the persons who are able to absorb knowledge to understand the background and context of the issues and who generate new knowledge and understanding.

It is a challenge for engineers, Susan Greenfield says, to help foster that creativity. "We have to think about ways technology can give meaning and significance to people rather than comfort and sensation. If people spend most of their time in front of a screen, we'll have to find out what the screens give them that they can't find in real life. We can then try to make something similar in the three-dimensional world. Or we can use the two-dimensional medium to deliver some of the things that are being lost. We may foster creativity, for example. I think it is necessary that engineers get more involved in game development. It's a very powerful and exciting technology, and we'll have to use it to give us a sense of greater stories. You can think about creating a second persona for someone—a corner on the Internet that is all for yourself and that nobody else can see. The opposite of Facebook, where you share your thoughts with 900 others. It would be more like a diary where you can be yourself, leave your thoughts and ideas, and keep your memories and photos—your own record for yourself. The computer can then help you to cross-reference it with novels or characters. That way, you would find comparisons on the Web and parallels that give extra meaning."

Greenfield also sees a remedy in the teaching of science. "Society has to become science literate. At the moment, science and technology are regarded as a minority activity. If people became more aware of how their brains function and how sensitive they are to their environment, they would have more respect for what they are doing. We know that once people are told how their brains work, they perform better in the classroom. It's empowering. Things like that would help people develop and get the most out of themselves. We have to do that now; otherwise, we will discover that we're suddenly in the world of the iPod generation, of people who text each other, who have speed dates, and who are living from moment to moment."

5.3

PROSPECTS OF CITIES

There's a greater than 50 percent chance that when you look through your window, what you see is a landscape of concrete, asphalt, and cars. More than half the world's population lives in cities, and the proportion is increasing—as are the problems associated with progressively denser and more aggregated communities. As we move further into the twenty-first century, the urban transition will gradually draw to a close after two centuries that transformed the human population from an agrarian society scattered over the surface of the earth to the highly compressed life of the city. The growth of urban living is one of the greatest paradoxes of our age. New technologies offer companies and individuals an unprecedented degree of locational freedom and mobility. We are increasingly able to see, hear, and sense one another, even when we are thousands of kilometers apart. More than ever people choose to live in close vicinity of each other, as if there were no other possibility to communicate.

Once most people live in cities the urban landscape will have become the dominant habitat for human beings and explosive urbanization will inevitably come to an end. We will then enter an era of posturbanization in which the city will have to find a new dynamic. Growth will no longer come by drawing people in from outside. Will cities maintain their scale? Or will urbanization go into reverse, turning downtown Shanghai, Mumbai, and Chicago into wastelands as the twenty-first century progresses? Detroit offers a glimpse of what happens when a city ceases to breathe. What used to be a theater is now a parking lot; the residual population grows vegetables on former city squares; empty office blocks gradually succumb to the weather; the car industry has collapsed, and nothing has emerged to replace it. How can we prevent cities from falling apart under their own weight?

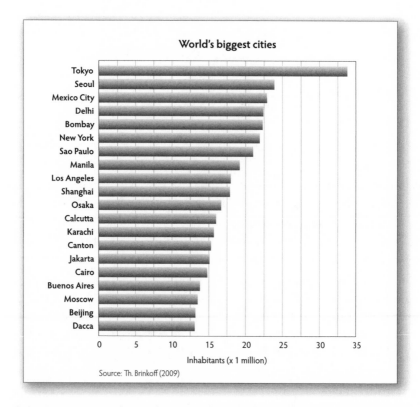

World's biggest cities

Inhabitants (x 1 million)

Source: Th. Brinkoff (2009)

A century ago, London and New York were the biggest cities on the planet. The most explosive growth nowadays can be found in the megacities of the developing world. Urban society has always been a catalyst for progress, but the burdens imposed by these ceaselessly growing agglomerations may soon begin to outweigh this particular benefit of city life. *Source*: Brinkhoff, T. (2009). *The principal agglomerations of the world,* http://vermeer.net/city

CITIES ARE ALIVE

Everywhere in a city, you'll probably hear the growling and snarling of an urban street. You feel its heartbeat. The city's hunger never wanes: It eats up its surroundings and excretes a constant flow of waste. It's a monster, an anthill, a dinosaur. Cities have been presented as living organisms for centuries, and it is a fitting metaphor. The statistics of large and small cities strongly recall those of biological organisms. Scaling laws apply to both; indeed, the rank-size distribution of cities is one of the best-known scaling relationships in human organization. It was already perplexing scientists in the early twentieth century, when only 10 percent of the world's population lived in towns.[1] There were only four cities with more than 2 million inhabitants in 1900, but even then, a scaling law applied to urban size. There were twice as many cities with around 1 million inhabitants as there were with 2 million. And the same is true if you compare towns with 100,000 and 200,000 inhabitants. Halve the size and you double the amount.[2]

Many more large cities exist nowadays, but the same scaling relationship continues to apply. We see something similar in animals. It's not surprising to learn that there are more mice than elephants. However, if you compare the statistics for the abundance of mammals with different body masses, you come up with a scaling law as clear as the one for cities.[3] It's not only in statistical terms that cities and animals resemble one another. The internal structure of the city is also very much like something from the animal kingdom. Cities and animals enjoy similar economies of scale: The larger the city, the less asphalt, power cabling, and shopping space you need per inhabitant. London, for instance, has fewer service stations per 10,000 inhabitants than Manchester does. Facilities like this can be used more efficiently in bigger cities. We find precisely the same with large animals; an elephant is twenty times the weight of a gorilla, yet its aorta is only three times as thick. That means the elephant uses its circulatory system more efficiently to fuel its body's cells. The larger the city, too, the more economically it uses its infrastructure. The same scaling laws apply whether we're talking about people traveling down roads or cells in the body linked by blood vessels and nerves.

THE PACE OF LIFE

There is one key respect, however, in which the metaphor of a living organism ceases to apply: Larger animals are slower. An elephant's heart beats more slowly than a gorilla's, and a mouse's heart flutters away faster than we can count. The whole metabolism in larger organisms is slower. A rat has to consume half its body weight in food each day, whereas we humans can get by on a far smaller proportion. In fact, the complete life cycle takes more time. Large animals need more time to mature, and they live longer, too. Back in 1932, Swiss physiologist Max Kleiber demonstrated this scaling of metabolic rates with body mass.[4] Around the turn of the century, it was shown that this is the result of the geometry of the internal infrastructure, the vascular systems of animals and plants. Big animals have no option other than to ration their consumption and to economize. They have less internal transport capacity available per living cell. Because of their larger volumes, they have fewer possibilities for getting rid of internal heat and waste. They have adapted to this by lowering their combustion rates and slowing down the pace at which they live.[5]

If it were economies of scale alone that drove the growth of cities, these would follow the same pattern as large animals, progressively slowing down and then stabilizing in size. But that's not what we observe. Larger cities clearly have a faster pulse rate; check out how quickly the people are walking past your window. Sociologists have measured the speed of urban pedestrians and found that people in downtown Tokyo, for instance, literally run to their destinations. Or at least that's how it seems to someone from a smaller city like New York or London. Tokyo people themselves see it as entirely normal. Meanwhile, someone from Pittsburgh might be surprised by the fast pace of New York.[6] It's not just a question of walking; every aspect of human life seems faster in larger cities. Thus, the organic metaphor doesn't offer a comprehensive description of urban life. In fact, it is turned on its head. If we want to understand how cities grow, we need to look for other—perhaps uniquely human—processes.

CITIES ARE ACCELERATING

Luís Bettencourt travels the world to study and experience the pace of life in cities. Born in Portugal, he studied physics in Germany, England, and the United States. He is currently a professor at the Los Alamos National Laboratory in New Mexico. Bettencourt's work takes in epidemiology, social networks, and urban dynamics. How large can a city grow? "In ancient times, Rome had about a million inhabitants. The writers of the time thought that a city with so many people couldn't exist for long," he says. "I recently visited Tokyo, and I felt a similar amazement. The city is almost twice as big as any other in the world. It is interesting to think about the mechanisms that allow a city of 35 million people to exist. It requires behavior and technology that enable that amount of people to work and live together.

The infrastructure is amazingly complicated. You almost don't live in geographic space anymore. When you change from one metro line to another, everything is dense with information. The underground network seems to have its own life. And it's so complex that even the locals get lost. So there are staff everywhere who can point you in the right direction. Life is conditioned by the place. It's absolutely fascinating." But Tokyo is not so very different, in fact, from Mumbai, New York, or Mexico City, which are about half its size but are subject to a similar pressure on space. "Some people think that cities are very different, but in fact, they are in many respects very equal. Their degree of similarity never ceases to amaze me. It's mostly their size that makes them different."

The similarity clearly appears from the urban data that Luís Bettencourt and collaborators have collected. They have uncovered two different scaling relationships that apply for a large number of cities in different parts of the world. One is the *law of economies of scale:* The larger the city, the more efficient its physical infrastructure. This is analogous to the organic scaling we discussed earlier. It applies to the amount of road surface, the length of electrical cabling, and the number of kilometers that cars travel. The other law is related to human activity and might be termed the *law of accelerated productivity:* If you double the size of a city, you get 15 percent more output per inhabitant. That goes for the city's gross domestic product but also—and more specifically—for its rate of patents and inventions. Wages scale in precisely the same way. Move to a bigger city and you'll earn more, but odds are you will also spend more. In fact, everything that constitutes the social fabric of a city scales as the law of accelerated productivity.[7] That includes the incidence of AIDS and violent crime, suggesting that walking isn't the only thing that speeds up as cities increase in size. "It seems that a city dweller does more in less time. The pace is higher," Bettencourt concludes. "As wealth creation accelerates, time becomes more valuable. You have to use your time more efficiently because the cost of living also increases. So acceleration is the central theme in cities. And it's clear that human interactions lie at the basis of it."

This is highlighted when you take a closer look at the statistics—crime rates, for example. "In larger U.S. cities, you have a greater chance of being confronted with violent crime like murder or aggravated assault. When you look at property crimes like burglary, however, there isn't much difference compared to smaller towns. In the United States, murder is often associated with organized crime and gangs, which have a social dynamic in which one crime leads to another. That's different from burglars, who tend to operate alone," Bettencourt observes. Both of the laws he and his collaborators have uncovered are valid for a large range of agglomerations. They are true of small towns but also of megacities. The law of economies of scale applies to the material fabric of a city and the law of accelerated productivity to human urban networks. Together, they capture the different forces that hold a city together and make it grow.

THE MECHANISMS OF GROWTH

From the early beginnings of sociology, scientists have observed how human relations change as a city grows. Social networks become more diverse, and labor begins to specialize. In recent times, Richard Florida, professor of urban studies at the University of Toronto, has received a lot of attention with his claim that high-tech workers, artists, and gay men foster a higher level of economic activity. According to Florida's hypothesis, these pioneers attract creative people along with businesses and capital.[8] It has always been very difficult to test such claims because sociologists could do little more than ask questions and count people. Hence, competing opinions exist among urban sociologists, not all of whom agree on the relative importance of this "creative class." Some argue that it is the presence of a large labor force in cities that enables specialization, and hence drives urban growth. Or perhaps it is the wealth of possibilities for consuming and for spending money in general. The micromechanisms that make a city grow are fiercely debated in the absence of empirical data capable of resolving the issue.

It has only recently become possible to monitor social interactions on a large scale. A great deal of sociological data is now becoming available because of the traces that human interaction tends to leave in modern communication systems. For the first time, it has become possible to monitor the behavior of many different people at once. Luís Bettencourt has been able to draw on mobile phone data from Kigali in Rwanda. "The data allow us to follow each individual in the region," he explains. "You can track, for example, a person who lives in a rural place with a lot of contacts in his close neighborhood. At some point, this person decides to migrate to the city. The pattern of contacts then changes. You can monitor how his network in the city develops at the expense of contacts with his native village. From the data, you can see that people with a rich network succeed in a city. Less connected people tend not to make it and return to their villages. We are currently analyzing these data to deduce how social interactions and social integration may spur urban growth."

Comparison of cities that stand out for either good or bad reasons provide further clues about urban growth mechanisms. The law of accelerated productivity holds for many cities. "But there are some that systematically underperform or overperform for their size," Bettencourt says. "The outliers often remain so for a very long period. I call it the 'local flavor' of the city. For example, cities in south Texas and inland California have done worse than others of their size for decades. By contrast, San Jose (Silicon Valley) and San Francisco have done very well from as far back as our income data go. They flourished 50 years ago, long before microelectronics or the dot-com boom. San Jose has doubled in size over the past half century, but it has maintained and even amplified its characteristics. Around 50 years ago, there was already an electronic industry south of San Francisco. It gradually developed into software. The seeds of growth that were sown long ago still make cities flourish. The opposite is also true. When there is nothing, there is no attraction; in fact, people tend to leave. So cities have a long memory. There is something persistent in the attractiveness of cities, something much more general than a creative class of Web designers. We are currently analyzing differences in employment among those cities, and this might give more insight into the mechanisms of growth."[9]

These processes are entirely familiar to any sociologist who has studied urbanization. We have now entered a period, however, in which we no longer need to rely on survey data and a few observations. The clustering of people can be observed today from massive bodies of data, and processes can be modeled in a computer. This kind of research could offer more clues regarding the forces underlying the law of accelerated productivity. Clues are offered by the fast-moving social dynamics expressed by mobile phone data and by the extended historical development of a particular local strength.

As more statistical data emerge, we will gain a clearer insight into the human interactions that make a city tick.

THE FUTURE OF MEGACITIES

The laws of economies of scale and of accelerated productivity also help us understand how a city grows—an extremely important factor when considering the questions we raised at the beginning of this chapter regarding the future of our cities. The two laws actually represent contradictory forces. The law of accelerated productivity makes it attractive to move to a city. As an individual, you earn more; as a company, you can achieve higher output. Growth attracts growth. Wealth, creativity, innovation, and other fruits of human interaction generate positive feedback that attracts people at an ever-increasing rate. The bigger the city, the more attractive it is—an idea borne out in our own time when worldwide urbanization is proceeding at an incredible pace. The law of economies of scale, on the other hand, restrains this growth. A bigger city has less infrastructure per inhabitant. That may be an advantage if everyone economizes on its use, like larger animals living a slower life, but the steadily increasing pace dictated by the law of accelerated productivity exerts growing pressure on the physical infrastructure of cities. As a city grows, it becomes impossible to widen its arteries without cutting deeply into the urban tissue. Far from slowing down, more and more people find themselves battling for their place in increasingly overcrowded streets and public transport lines.

Viewed in this light, it is amazing that a city manages to hold itself together and keep growing. It's easier to make new arrangements in a small town than it is in a big one, making it logical to quit the city if there is nothing there to bind you. Yet the attraction of accelerated productivity is evidently so high that people take for granted the constantly congested and overloaded arteries. A compromise is the suburbs: People can live at night at relatively low densities and during the day benefit from the concentration provided by the city. Yet this further stresses the transportation systems. It is the attraction of the human fabric, then, that binds the city together. In the long run, this is not sustainable. How much faster can it become? In certain cities, the pace of life and the pressure on the infrastructure are already close to unbearable. Workers have to adjust to this fast rhythm. Japanese has a special word for someone who works himself to death (*kar shi*). We resort to personal coaches and yoga masters to help us get through urban life. Will cities grow themselves into a coronary? There is, after all, an upper limit on how fast pedestrians can run. The point will be reached at which the urban organism outpaces the human. At what point does the repulsion exerted by an overcrowded space become so strong that it causes a city to collapse?

That brings us to the central question of this chapter: How can we keep our cities habitable? "Cities avoid a crisis by reinventing themselves," Luís Bettencourt thinks. "They change the way they operate. Innovations are not only used to make new products but also to change the urban infrastructure and the way people live and work." The largest cities were the first to build underground railway lines and expressways, and labor patterns changed to adapt to the pressure of the urban environment. "Cities need innovation to overcome the constraints of the urban fabric," Bettencourt confirms, citing New York as an example. "Over the last 200 years, the population almost always kept growing but not at a constant rate. Growth surged on the arrival of immigration, the textile industry, and other manufacturing in the early twentieth century and on mass media and the dotcom activity in the '90s. When growth was too fast for the internal dynamics of the city, you always saw a change. The city couldn't have sustained its industry without technology to clean the smoke from its chimneys or reduce overcrowding. When crime exploded because of the advent of so many new inhabitants in the 1970s, new approaches to law enforcement eventually reduced the nuisance to an acceptable level. So innovation keeps the city together and keeps it changing and growing. And there is a pattern in those innovations. To sustain its growth, the city should have an ever-increasing pace of innovation. The larger the population, the shorter the time to the next crisis and the next innovation to overcome it. People in New York and other large cities today will experience several major changes in their lifetime."

Bettencourt saw this pattern in the historical data from New York. But he also deduced it from the mathematics of the city's attractive and repulsive forces. "It's an unstable dynamics. You're on a treadmill and have to make one change after another to hold the city together. Constant action is needed to ensure that the good things about a city keep dominating the bad things. Otherwise, you will face a collapse," something that has already happened with cities like Buffalo, Pittsburgh, and Cleveland, which have shrunk since 1960. They failed to discover the next innovation cycle, allowing negative feedback to take hold and to repel people from the city.

Is there nothing we can do to turn the largest cities away from their accelerating pattern of growth? "We are only beginning to understand the mechanisms behind this growth," Bettencourt admits. "Much more theory and modeling and many more data are necessary. But it ultimately comes down to the question of whether we can stop the acceleration of the pace of growth. We will continue to innovate: That's part of being human. Stopping innovation would be against our nature. But it is the relentless acceleration that has to stop. Maybe one day, when technology makes possible concentration without congestion. Maybe even physical proximity is no longer necessary to sustain accelerated productivity. We will look back at the beginning of the twenty-first century as the golden age of cities, tinged by some of the greatest challenges our species has ever faced and at the same time marked by some of its greatest achievements."

TOWARD SUSTAINABLE CITIES

A lot of computer simulation and data crunching is going on to better understand the human fabric of cities. The high pace of innovation clearly doesn't resemble anything in the animal kingdom. If cities were like living organisms, they would have been abandoned already. The organic growth of their physical infrastructure might resemble the patterns we see in nature, but the human interaction does not. The findings of Luís Bettencourt and his coworkers remind us of the many networks we have encountered throughout this book. The solutions identified in those other cases could, therefore, be helpful in contemplating the future of our cities, too. Communication networks display certain similarities with the human fabric of cities, as the exchange of information is what determines the law of accelerated productivity. These communication networks form what urban sociologist Manuel Castells calls the "space of flows," the information exchange used in the real-time, long-distance coordination of the economy.[10] Picture the city as a node in the global Internet. Rather than the humming computers in a data center, there are individual people who exchange ideas, thoughts, and knowledge. There are striking similarities with the Internet hubs that collect, process, and alter data flows. The size distribution of hubs in the Internet follows a scaling law much like that for cities. There are many parallels between urban expansion and the evolution of communication networks. We saw in chapter 3.2 how Internet hubs—like our cities—are growing extremely rapidly. The processing of information at these hubs is increasing faster than the number of Internet access points, just as productivity in cities grows more rapidly than the number of people in them.

It is interesting to see how computers cope with the increasing pace of communication. The traditional solution in microelectronics has been to increase processors' clock speeds and become more compact to reduce interaction times. That's just like cities that tick faster as they become larger. Microchips also have made improved use of the third dimension, again just like cities. As we saw in chapter 3.1, this development in microelectronics stalled around the turn of the millennium when the pace had become so high that it caused chips to overheat. The physical limits of the material had been reached. We are likewise reaching the physical boundaries of the human capacity to function in cities. It will be hard to walk much faster than people in Tokyo do already. Processors could still increase their speed by using multiple operational cores working in parallel. That would be like cities with multiple centers or like neighboring cities that are closely integrated and distribute work between them. In microelectronics, it generally doesn't pay to have more than eight parallel cores because the accelerated complexity of interactions tends to nullify the gains.

Hence, there are limits to what you can do at the largest Internet hubs, where further growth is becoming increasingly difficult. One consequence is that smaller hubs are gaining in importance. The strongest growth is no longer in the data centers of Tokyo, London, or New York but at the second tier of hubs, which are establishing direct mutual links that avoid the biggest hubs. After all, these data centers have all the necessary expertise and facilities, and additional communication lines are easy to obtain. Only computer services where every millisecond of interaction time counts will continue to use the biggest hubs. This development is why network links within the Internet have been growing denser since around 2001[11] to bypass the largest hubs.[12] In a similar manner, we might expect the largest cities to specialize in highly interactive tasks, which is precisely what urban geographers like Saskia Sassen have observed.[13] Some of these "megacities" have been respecializing in businesses that need the proximity of a high-paced, well-connected urban agglomeration. Sassen focuses in particular on high finance, with other activities increasingly looking elsewhere—something that will ultimately limit the growth of the largest cities. What's more, reliance on a few high-paced sectors makes megacities vulnerable. It remains to be seen, for instance, how they will come through the banking crisis that began in 2008.

AVOIDING COLLAPSE

The development of computer centers in the 1980s witnessed a more extreme outcome. As computers grew increasingly powerful during the preceding decade, they moved into centralized computer centers. These allowed expertise and expensive resources to be shared among several users, outweighing the inconvenience of remote access and the complicated scheduling of computational tasks. This is once again reminiscent of cities, where accelerated productivity makes us tolerate the nuisance of having everything cramped together. The pattern came to an abrupt end with the advent of the personal computer and cheap networking, which sparked the sudden collapse of these computer centers. One after another was dismantled, leaving their climate-controlled rooms as bereft as downtown Detroit. It's entirely possible that large cities will similarly fragment into smaller PC-like communities with strong mutual connections.

Twenty years have passed since the advent of the PC—an eternity in computing history. Scientists are now working hard to make over the biggest Internet hubs in the way we described in chapter 3.2. Electronics there could be replaced by devices that directly manipulate the light pulses carried by the fiber-optic cables that form the spokes of the Internet. This has the potential to speed up interaction and reduce the amount of heat that is generated. The corresponding development in cities would be to enhance people's ability to interact without the problems this currently entails—not by somehow magically altering human nature but by boosting our performance. It could be done, for instance, by providing tools to take over some of the work. The routine part of human interactions could be left to the machines, freeing us from all sorts of administrative tasks and placing greater emphasis on our capacity and creativity.

This wouldn't reduce the need to see each other. On the contrary, personal interaction needs to be intense and close at hand. For the time being, that still requires us to live in close proximity with one another. In the future, by contrast, broadband communication facilities could enable us to socialize virtually with other humans in real time and at real size, allowing us to "surround" ourselves with other people even though they might be physically located far away.

New communication capabilities simultaneously increase the complexity of projects and the pace at which they can be undertaken, which probably explains why the age of urban growth has coincided with the rise of the information age. The computer network metaphor offers an impression of how our cities might evolve toward greater stability and less explosive growth but also toward collapse. The network approach clearly has something to teach us about the future of cities, which is why it is already being intensively pursued.[14]

5.4

DISASTER SCENARIOS

A hurricane striking the Chinese coast is ten times as lethal as one hitting the United States.[1] The number of U.S. victims is limited because of better precautions, warning systems, and evacuation methods. More effective observation and communication can save lives. A century ago, hurricanes killed around 7,000 Americans every year, whereas nowadays there are only very few hurricanes of the lethality of Katrina.

That progress has yet to reach every corner of Earth, says Guus Berkhout regretfully. This Dutch geophysicist has immersed himself in the mechanisms of disasters and disaster prevention since the beginning of his scientific career—first as professor of seismic imaging and later as professor of innovation at Delft University of Technology in the Netherlands. We talked to him at the university campus that lies 3 meters below sea level. At his laboratory, Berkhout analyzes the early warning systems and contingency plans that will be needed to protect both his lab and his compatriots. "We can't stop earthquakes, volcanic eruptions, hurricanes, or tidal waves from happening," he stresses. "And we may never be able to predict hurricanes or earthquakes with sufficient accuracy. Nor can we hope to prevent people from living in dangerous places. They are simply too attractive."

Human beings indeed seem addicted to living on the edge of catastrophe. The World Bank has calculated that a fifth of all countries are under permanent threat of natural disaster, with some 3.4 billion people—roughly half the world's population—at heightened risk of being killed by one.[2] Yet unsafe regions are often exceptionally popular places to live and work, one reason being that floodplains and the slopes of volcanoes are highly fertile. The climate is milder along the coast, the soil better, and transport more efficient than farther inland. Even the likelihood of earthquakes isn't enough to persuade people to live elsewhere, as witnessed by some of the most densely populated areas of California and Japan. Current migration trends—moving to where the action is—suggest that the proportion of people living in unsafe areas will only increase. The immediate benefits of living in a particular place often seem to outweigh the vague likelihood of some future disaster. This is why so many people are apparently willing to put themselves in harm's way.

WE SAW IT COMING

"The main things we can do are early detection and mitigation of the effects so that people have a greater chance of escaping with their lives," Guus Berkhout thinks. "I envisage a global system of smart sensors and responders; you need to know precisely what's going on—from space, on the earth's surface, and beneath it. All that data then have to be combined and acted on."

The necessary technology already exists. Berkhout mentions the rapid detection that is routinely carried out on behalf of the Comprehensive Nuclear Test Ban Treaty Organization (CNTBTO) in Vienna. The CNTBTO manages a global network of ultrasonic measuring equipment designed to detect nuclear tests. The system is so sensitive that it also registers when fishermen use dynamite to boost their catch or when part of an iceberg breaks off. Monitors instantly picked up the earthquake that unleashed the tsunami on December 26, 2004. But the organization was only mandated to monitor for nuclear tests and was therefore unable to warn Indonesia, Sri Lanka, or India of the tidal waves that subsequently drowned 230,000 people.[3]

The activity of the CNTBTO monitoring system has since been broadened so that in the future it will be able to issue tsunami warnings, too. But that's only the beginning, Guus Berkhout believes. "We should extend the technology to pick up signals of other imminent catastrophes as well." He points to the example of Japan, where a system has been set up that exploits the fact that tremors take time to travel through Earth's crust. Provided you're not at the epicenter, you have a few seconds to get out of a building. When an earthquake struck the Japanese city of Sendai in 2005, the 16-second warning was enough to successfully evacuate a school.[4] All this naturally requires a considerable amount of discipline plus incredibly fast communication.

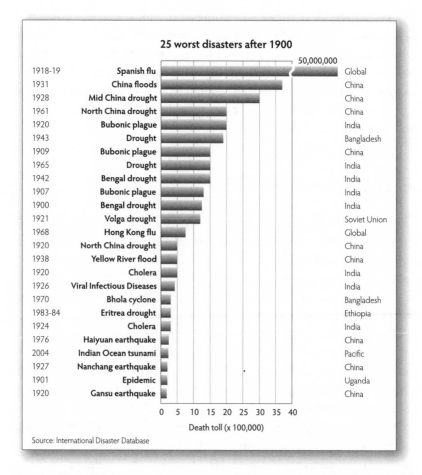

25 worst disasters after 1900

Year	Disaster	Location
1918-19	Spanish flu	Global
1931	China floods	China
1928	Mid China drought	China
1961	North China drought	China
1920	Bubonic plague	India
1943	Drought	Bangladesh
1909	Bubonic plague	China
1965	Drought	India
1942	Bengal drought	India
1907	Bubonic plague	India
1900	Bengal drought	India
1921	Volga drought	Soviet Union
1968	Hong Kong flu	Global
1920	North China drought	China
1938	Yellow River flood	China
1920	Cholera	India
1926	Viral Infectious Diseases	India
1970	Bhola cyclone	Bangladesh
1983-84	Eritrea drought	Ethiopia
1924	Cholera	India
1976	Haiyuan earthquake	China
2004	Indian Ocean tsunami	Pacific
1927	Nanchang earthquake	China
1901	Epidemic	Uganda
1920	Gansu earthquake	China

Death toll (x 100,000)

Source: International Disaster Database

The deadliest disasters continue to occur in the poorest parts of the world. Wealth and disaster preparedness go hand in hand. Our challenge is to install reliable global disaster-warning systems and to help local communities improve their own safety. *Source: The international disaster database* (2009), http://vermeer.net/disaster

Satellites could help provide early warnings by observing the electrical phenomena that immediately precede an earthquake. That would give people a few additional seconds.

Berkhout continues: "Nowadays, we increasingly combine monitoring data with scientific models to predict the near future. That capability gives us more time to take emergency measures prior to 'time zero.' Actually, scientists more and more realize that the ultimate test of their models is not fitting past measurements—I call that history matching—but predicting future events. A topical and relevant example is climate change. The prediction power of today's climate models is yet insufficient, causing continuous updating of the prediction results in the history matching phase. This problem plays an important role in the scientific debate about climate change."

ORGANIZATIONAL INERTIA

The deployment of more sensors and better models would undoubtedly help fill in the gaps in our intelligence. Berkhout argues, however, transforming all available information into actions is even more important. "Rapid response isn't only a question of early warning by improved prediction. When it comes to action, information tends to get fragmented in the process. It's split between different regions and organizations. It's difficult to exchange information because of incompatibility between the different infrastructures. And there's never a proper agreement about who bears overall responsibility in the event of a disaster. Even if you could collate all the information, there would still be problems communicating your decisions down the chain of command in time. The decision-making hierarchy as it stands is not only too slow; it's also too vulnerable. Because of its linearity, the failure of one part jeopardizes the total chain. The robustness of linear chains is notoriously poor."

That much was painfully obvious in the aftermath of the 2004 tsunami. A Thai meteorologist who had access to up-to-date information about the earthquake decided not to sound the alarm because he didn't believe the signals were sufficiently clear-cut. A top-down, linear system can thus fail due to a single decision. Decision-making processes are therefore crucial, also in the case of the CNTBTO. That's also true for the coordination of relief actions. Despite having plenty of warning, disaster relief in New Orleans was chaotic because the hierarchical command structure prevented the right decisions from being taken. "You could see the whole thing grinding to a halt," Berkhout says. "It simply doesn't help to have people obediently executing whatever prior instructions they happen to have been given."

Recent studies show that a storm surge in the Netherlands could create many thousands of victims due to organizational inertia.[5] The Netherlands is a low-lying country that has defined itself since the

Middle Ages through its constant struggle with the sea. It now realizes that it risks losing that battle because of the ever-increasing complexity of a society in which decisions and evacuations take too much time to execute. Guus Berkhout's own lab is located in the danger zone.

SELF-ORGANIZATION

"Current contingency plans still feature a linear and laborious process that follows strict protocols," Berkhout continues. "The people involved don't have all the information they need to improvise a decision in response to something unforeseen. In New Orleans, for instance, someone in the chain of command decided to empty a supermarket that was going to be submerged so that the supplies could be taken to a reception center. A useful idea, but should it have been the priority? You also suddenly had a situation where soldiers who were supposed to be filling sandbags were switched to other duties, such as fishing animals out of the water. Once again, a useful thing to do, but no one knew whether it was the most pressing task at that given moment."

Hierarchical networks only work when there are few surprises and few failures.[6] But surprises and failure are precisely what happens during a catastrophe. "Hierarchical orders by centralized command inevitably arrive too late—or not at all—when it comes to disaster relief," Berkhout explains. Redundant information links are therefore essential to compensate for the lost information. The optimum organizational structure is not mechanistic: Similar to innovation,[7] a decentralized, organic network is the most robust when it comes to coping with unexpected events and missing data. "You have to make sure that ordinary people can make decisions themselves when a crisis breaks out. Crisis management ought to be largely self-organizing and flexible. The chain of command should only concern itself with resource management and strategic decisions. In other words, it is a balance of 'situation awareness' at the site and 'total overview' at the command center. That implies a major change in the kind of tools and organization you need, not to mention a huge cultural shift."

Modern information and communication technology, Guus Berkhout believes, accelerates the cultural shift toward a high degree of self-organization. Experiments are already being carried out with displays in police cars showing where fellow officers are located. It resembles online social networks like Facebook, Twitter, and LinkedIn and allows users to find people in their neighborhood. The control room no longer has to decide who's closest to an incident; the patrols can do it themselves with the assistance of their onboard computer. Likewise, when disaster threatens, people can make their own way to an evacuation center, provided they have sufficient information. "We need less money and people for that, and it is much faster," Berkhout says. "Self-organizing collaboration techniques are sufficiently developed now for us to put them into practice. That would be a real breakthrough in terms of safety. It's urgently needed, too, because society is constructing more complexity all the time. That means increased inertia unless more people have access to information and can decide about their own emergency relief."

LET EVERYBODY KNOW

Leaving the detail of evacuation to the people themselves obviously has its potential dangers. All sorts of unexpected and powerful interactions might develop. People might begin to run after each other as lemmings. As in any other complex dynamical system, the situation could theoretically become unstable and even chaotic. "That wouldn't happen as long as macrodecisions are still being made at the strategic level," Guus Berkhout says firmly. "What's more, local decisions don't all have to be rational, provided that the majority of them are. Intensive contact will make sure that everyone falls into step. It's the same kind of adjustment you see on an expressway, where everyone decides how they're going to drive. Despite the high speeds, cars can travel safely by constantly making minor adjustments relative to one another. It's a phenomenon called *intelligent swarming*."

Surely, swarms could also take a wrong turn. Shouldn't we be afraid of the way the media might whip people up in the face of imminent disaster, causing everyone to rush off in the same direction like lemmings? Not according to Berkhout: "The media only behave that way when there isn't enough real information available. If lemmings knew where they were headed, they wouldn't jump into an abyss. If everyone has access to the same information, there's no room left for speculation. Journalists would no longer have to base their reports on the statements of a harassed spokesperson. People could see for themselves what the estimated casualty figures were based on. We're already seeing with blog culture how people are starting to act as their own journalists. We're beginning to understand that you have to involve the media and the public in crisis management. They're part of the information chain."

Berkhout emphasizes: "All these important aspects are now being investigated with the field of serious gaming. We are entering the petaflop era in digital computing, enabling us to simulate all kinds of disasters in a very realistic manner. Next, it allows us to evaluate how the alternative decision-making processes work out in the different scenarios. This is the way we should prepare ourselves for future disasters."

Disasters will still be with us 20 years from now; that much is certain. But we can mitigate their consequences so that catastrophes have fewer victims, according to Guus Berkhout. "Better scientific models, better prediction technology, and a fundamentally different kind of organization, all three will be needed if we want to limit casualties and damage. It is a moral obligation for all to make that available with high priority."

5.5

RELIABLE FINANCE

As we were drafting our first version of this chapter, the world was abruptly seized by the worst economic crisis since the 1930s. Having just written about the impact that instability, the bonus culture, and the bursting of financial bubbles might have on our collective future, it was disconcerting to see those ideas leap from the page and run amok in the global economy. Rather than tempt fate any further, we put our text on hold and went back to Jean-Philippe Bouchaud, the financial expert with whom we had discussed the potential for precisely this kind of ominous development a few months earlier.

Bouchaud knows just how fast money can move. He has set up his computer systems in three separate continents, as close to the major financial centers as possible, because communication between the continents can lag by a few milliseconds—a costly delay he simply can't afford. "The speed of hot money is close to the speed of light," he jokes. "It's relativistic finance." As a physicist, Bouchaud is well aware of the constraints that the theory of relativity imposes on our actions. But that's not the only inspiration he has drawn from the laws of nature. Bouchaud's focus is on the most refined physics, which uses the behavior of individual atoms to explain how collective phenomena such as electrical conductivity and magnetism arise. Nowadays, he's professor at the prestigious École Polytechnique in Paris, but he has been applying his knowledge of collective phenomena to financial market prices for many years now. Together with Jean-Pierre Aguilar and Marc Potters, Bouchaud is the cofounder of Capital Fund Management, which rapidly grew into France's largest and most successful hedge fund. What makes the fund so successful is, perhaps, that Bouchaud's ideas differ fundamentally from the standard approach that economists have developed over the years.

THE END OF ECONOMICS AS WE KNOW IT

A huge amount of research was carried out in the 1950s and 1960s to identify patterns in financial markets. This gave rise to the "quantitative" economic theories that banks and financial institutions now use routinely. The associated mathematical models enable traders to analyze the markets and computers to trade automatically. Yet the success of these economic theories has so far proved disappointing, Jean-Philippe Bouchaud notes. "What has economics achieved? Time and again, economists have been unable to predict or avert crises, including the 2007–2008 worldwide credit crunch." Model makers were shocked in 2007 when the financial markets began to display behavior that, according to economic orthodoxy, should have occurred less than once in a billion years.[1] Prices moved in precisely the opposite direction from what the theory predicted. Investors who believed they had hedged their risks watched their assets evaporate. Computer programs that used economic theories to trade automatically were suddenly worthless. A snowball effect turned into an out-and-out avalanche, rapidly driving down prices. This is behavior characteristic of an unstable system that has moved far out of kilter, crossing a critical threshold and then basically running out of control.

Standard theory assumes that economic agents make effective use of all the available information. Experts are thought to know what a business is worth and hence also what a share of it should cost. In theory, the price is always rational: The value of a business rises and falls as news about it is assimilated. Market trading is triggered until prices are brought back into balance. The entire modern economy is based on this theoretical equilibrium. Scottish economist Adam Smith claimed as early as 1776 that there was "an invisible hand" balancing supply and demand with prices invariably reflecting true values.[2] That makes sense: Why would investors pay more for a business than it is actually worth and risk losing their capital? "If the 2008 crisis tells us anything," Bouchaud retorts, "it is that a lot of agents acted irrationally. It's wrong to say that people make decisions in isolation based on their own judgement and their own rationality. In hindsight, the idea of a free market's omniscience and perfect efficacy looks more like anticommunist propaganda than credible science. In many cases, it isn't government regulation that drives a market out of equilibrium as the proponents of the free market would have us believe. More often than not, it's the traders themselves. People mimic each other; they get scared or overconfident. Traders are absolutely not a well-informed body of professionals who, between them, bring about the right balance." This reminds us of the thoughts presented by Susan Greenfield (chapter 5.2). Probably, most of the traders have a lot of knowledge but an insufficient understanding of the relevant issues. Like most gamers, they can react quickly on events and observations without having much interest in the context.

Bouchaud believes we need a new economic theory—one that takes account of market imbalances—if we want to properly understand the mechanisms that underlie trading. "We need a theory that can model

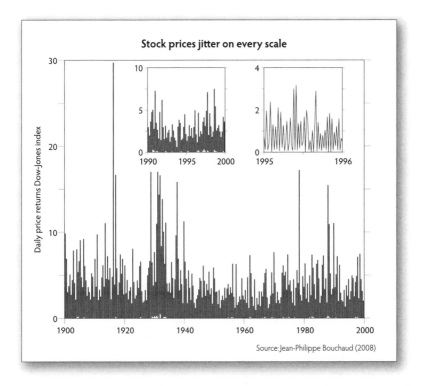

The pattern of financial trading displays a similar structure whatever timescale we choose. Research shows that very little correlation exists between trading and business news. Traders seemingly have a mechanism of their own. The challenge facing us is to deter the kind of herd behavior that triggers financial market bubbles and crashes. *Source*: Jean-Philippe Bouchaud (2008). Personal communication.

the irrationality of traders and consumers—a theory that includes essential elements of how *Homo economicus* behaves. We don't have a theory like that which is surprising because it seems so obvious that there are processes driving prices out of equilibrium."[3, 4]

MARKETS DRIVEN BY FEAR AND GREED

One of the most urgent things we need to do, Jean-Philippe Bouchaud thinks, is to incorporate herding behavior into the models. "We learn by imitation, which is probably a good survival strategy. But mimicry has its dangerous side, too. It can give rise to collective phenomena that are difficult to control. I've studied simple situations of mimicry, like applause. People in a concert hall are sensitive to other people applauding. You might expect applause to die out smoothly. But because people are listening to what everyone else is doing, it actually stops abruptly. Nobody wants to be the last person applauding. Mimicry can cause sudden changes. If the people in the audience couldn't hear one another, their applause would fade away gradually." Another example of herding behavior is addressed in the chapter on disaster issues (chapter 5.4). Running away when others do is probably a correct survival approach, but it can also cause mass movements going out of control for the wrong reasons.

It's not hard to apply these ideas to financial markets. Suppose traders don't have firm market insights of their own but are influenced at least in part by the opinions around them. It's not an unreasonable assumption, given the complexity of these markets. Traders would therefore tend to be optimistic when others are, too, encouraging collective behavior reminiscent of a surge in applause and ultimately causing the market to boom. That's precisely what happened in the years leading up to the 2007–2008 credit crisis, when collective euphoria dampened people's sensitivity to several negative portents that could be discerned in the markets. By contrast, the reversal of the upward trend came abruptly as high expectations turned out to be unsustainable and the new, grim prognoses cascaded through the minds of traders. The applause came to a sudden stop. Per Bak, a pioneer in the science of nonlinear dynamical systems, has explored this process of mimicry quantitatively. He has found that a simple model based on human imitation captures the odd patterns described by stock market prices—a phenomenon that has puzzled economists for decades.[5]

THE PROBLEM WITH PRICE FLUCTUATIONS

Financial market prices jump around for no obvious reason—something economists refer to as *volatility*. "The level of market activity is much greater than you would expect in a rational setting," Jean-Philippe Bouchaud explains. "Individual stocks typically vary by 2 percent a day. But how can the value of a company change so much, so frequently, even on days when there is no news affecting it? Stock prices even fluctuate sharply from second to second—movements that are much faster and more frequent than any news that might be feeding through. So there's no one-to-one relationship between price movements and news about the companies in question," Bouchaud says firmly. And he also proved that quantitatively.[6]

That's not the only problem caused by price movements. Crashes occur much more frequently than they should according to standard economic theory. No one appears to have taken seriously the actual numbers associated with past crashes.[7] The statistics of financial market fluctuations do not resemble random movements around an equilibrium point. They are more reminiscent of the patterns associated with those archetypal examples of instability—earthquakes and avalanches. It is a power law of the kind we encounter throughout this book, which in turn signifies a process in which tensions build up before suddenly being released. In processes of that kind, major fluctuations are far more frequent than equilibriums.

Per Bak's calculations suggest that these fluctuations can be readily explained in terms of herd behavior. That same behavior shows why trading continues even if there is no news on which to base it. Bak's herd model may be grossly oversimplified, but it illustrates that certain simple assumptions are sufficient to understand phenomena that standard economics cannot explain. To make the models more realistic, other irrational factors should also be included, such as the rewards that encourage traders to take risks. Bouchaud argues that the tendency to gamble is an inherent human trait. "Traders are given incentives to gamble and to take risks. The worst that can happen is that they lose their job, while the best outcome is that they receive a life-changing bonus. This asymmetry in rewards encourages people to take risks that are beyond anyone's control. The positive feedback it induces severely disrupted the markets in the 2008 crisis."

The science of nonlinear dynamics has therefore inspired new models that capture certain important and poorly understood aspects of the markets. "But the picture is still patchy," Bouchaud warns. "This is research on the fringe of economics: We're still outsiders. I receive more invitations to give talks these days, but these ideas haven't percolated into mainstream economics. There are no textbooks, and it isn't taught at the university. That means the old beliefs are still being perpetuated."

MAKE MARKETS MORE TRANSPARENT

The fact that stock market movements match the statistics of nonequilibrium processes so neatly is an undoubted success for the econophysicists. It is hardly reassuring, however, to think that the violent swings and the bubbles and crashes that accompany them are an innate characteristic of our financial markets. The only hope is that deeper insight into the underlying processes might furnish us with tools capable of dampening these fluctuations and preventing the next crisis. One of the first things regulators should do, Jean-Philippe Bouchaud thinks, is to make the markets more transparent. "Without data, you can't hope to understand the markets. But making markets more transparent is very difficult, as the banks strongly oppose it. They have an advantage in the marketplace when they're able to conceal something. It's like when you sell a car; you don't necessarily want the buyer to know about every little defect. But efficient markets require information to be known to everybody. Otherwise, the buyer will never pay a price for the car that reflects its true condition."

Making markets transparent is also difficult because financial networks are constantly evolving. "Financial institutions innovate like any other industry," Bouchaud says. "New products are often not understood that well at first, which creates new risks that are barely regulated. That's amazing. For food and pharmaceuticals, there are regulatory bodies that require you to assess the risks before you're allowed to bring new products to market and rightly so because they may endanger large parts of the population. As we have just seen, however, a financial crisis can hit ordinary people very hard, too. We really ought to regulate new financial products as well and require that all the market data are made public." Once you have the information, you can use computers to analyze it and test financial models. "The computer can analyze lots of data instantaneously. If you detect precursors of instability, you can then counteract them immediately. Feedback loops are much faster with computers than with human decision making. They enable swift action to be taken, which may allow for a soft landing when things get out of control."

A research group in Zurich is putting that idea into practice. International earthquake expert Didier Sornette has set up a "Financial Crisis Observatory" at the Swiss Federal Institute of Technology (ETH), where his computers look for simple patterns signifying irrational herd behavior in much the same way as they previously detected signals of stress in Earth's crust. Any quantity that increases at a faster than exponential rate is suspicious, Sornette believes. Such behavior suggests overconfidence, with people following one another in what has become an unsustainable upward movement on the part of the market. Housing prices in the United States behaved in precisely this way in the run-up to the 2008 financial crisis, and similar patterns were detected during the Internet bubble in the 1990s and the 2006–2008 oil price bubble. "This is the bare minimum that every investor should do," Bouchaud says. "We did it with our investment fund when we noticed that credit default swaps at Lehman Brothers were growing at a dangerous pace. We pulled our money out a few days before the bankruptcy. It's absolutely necessary to make these kinds of analyses, but you can only do it if the data are there."

STRIPPING OUT THE IRRATIONALITY

The other requirement, Jean-Philippe Bouchaud argues, would be to take the human emotion out of the markets. "Informed people trade less because they know there really is no new information affecting the value of their investments beyond transaction costs. Trading in and out too frequently is often a sign that people haven't done their math properly. Using automated trading would flatten out unnecessary trading. Computers have no emotions, so automatic trading would overcome the problem of overconfidence and hubris. Computers don't receive bonuses, they don't make crazy decisions, and they don't suddenly change their minds. They only do what they have to."

Computer trading might, however, introduce other sources of risk, such as bugs and errors, Bouchaud cautions. "If computers use algorithms that are based on incorrect economic ideas, they might inadvertently introduce positive feedbacks, as has happened several times on the financial markets. So you have to be aware of their limitations. Properly applied, though, using computers could stabilize markets in the longer run. Market volatility has declined over the last 5 years before the crisis, which has partly been caused by automated trading strategies."

Most important, however, the application of nonlinear dynamics to economics might give us a better understanding of market mechanisms. Computer simulations and other new network science tools could provide a deeper insight into interactions within a trading network, which might in turn teach us novel ways of structuring the markets for greater stability. One possible solution might lie in making markets more local. That's certainly a useful strategy when a dangerous illness breaks out, and it closely resembles the way ecosystems tend to stabilize. Perhaps we need to introduce friction to the market in the form of taxes on financial transactions. That would also reduce global dependencies. This is an idea that Sweden now wants to introduce. Or maybe there is some as yet unthought-of measure capable of relieving the markets. "It's very important to develop better models that include nonequilibrium processes," Bouchaud stresses. "An accurate model—even if it's not yet particularly sophisticated—would give us greater insight. That would be a lot better than elaborate models based on the wrong assumptions. Biased models give you the wrong ideas about market mechanisms. A different scientific culture has to emerge. As physicists, we have grown up in a culture where you learn to doubt as much as you can. Economics is based on blind belief in a few assumptions. That has to change."

The performance of Capital Fund Management, of which Jean-Philippe Bouchaud is a director, illustrates the power of a more rational and transparent approach to financial trading. "In 2008, when most funds were losing value because of the crisis, we realized a profit of 8.5 percent. That's because we've always stayed away from toxic financial products. We only trade in financial products on markets that are fully transparent and where all data are available. Our strategy has always been to trade at a constant risk. In other words, if volatility—and hence, risk—increases in one market, we pull out a little to compensate. It's a completely rational strategy, and one that clearly paid off. The only drawback was that our customers started panicking: When the crisis was at its height, a third of our customers pulled out of the fund. Not because of our performance but because they needed to cover losses elsewhere or out of sheer panic."

5.6

PEACE

A million people die every year as a result of war and terrorism.[1] According to these statistics, armed conflict will cost another 20 million lives in the coming two decades. Is there anything we can do to stop that from happening? The origin of war is one of the oldest questions of humanity. Every major religion sets rules that limit armed conflicts. Yet war seems to be a destructive power that is present throughout history. Is it within our power to prevent war and terrorism?

British meteorologist Lewis Fry Richardson was one of the first to apply statistical analysis to warfare. Richardson was a Quaker whose beliefs prevented him from serving in combat, so he drove an ambulance during World War I instead. It was then that he first began to collect data on the death toll attributable to armed conflict. Richardson went on to study military confrontations from 1820 to 1945, ranging from minor local skirmishes to all-out world wars. As we might expect, he found that the deadlier the conflict, the less frequent its occurrence. What was not expected, however, was his observation that the frequency of wars follows a similar kind of scaling law as earthquakes and avalanches. There are roughly fifteen conflicts each century costing more than a million lives. Those with a death toll above 100,000 occur 100 times, those with 10,000 or more deaths 800 times, and so on. A tenfold increase in lethality thus corresponds with an eightfold decline in frequency.

This came as a great surprise because it suggests that wars don't occur randomly. The fixed proportion of smaller and bigger conflicts shows that they are interrelated and that there is a common set of forces driving a dispute toward war. This is a profoundly worrying conclusion for historians, who customarily ascribe each new outbreak of violence to a set of unique contingencies. Yet this is only part of the story. Scaling law statistics suggest that there is something universal shared by all wars—something that may be inherent to human society. Sooner or later, another führer will emerge somewhere around the world, unleashing another war with its own timing and outcome. If large-scale conflicts are indeed recurring events, we need to start worrying about the next conflict that will cause more than a million deaths. This prompts a whole series of disquieting questions. Why hasn't humanity learned anything from earlier conflicts? Why doesn't each war bring us any closer to lasting peace? Can we alter our seemingly bloody destiny? Are we really just following the laws of nature, which dictate that armed conflict will inevitably follow the same pattern as earthquakes and avalanches?

THE WEB OF PEACE

It is indeed tempting to interpret the statistics of war in this kind of deterministic way. If we focus on longer periods, however, we find that the scaling laws are by no means constant. The eighteenth century, for example, was full of deadly conflicts, whereas the nineteenth was relatively peaceful.[2] Changes in global politics have altered the nature of conflict, and it's also entirely logical that the scale of wars should vary over history. Even without any shift in human nature or the relative belligerence of our societies, technological advances strongly influence the outcome of diplomacy and the scale of killing. The advent of the railways in Europe, for instance, changed the size and reach of armies.[3] The ability to move troops from one border to another in the space of a day underpinned the foundation of the German empire in 1871. Railways also changed the structure of industry and hence the complexity of society. And that's just one example of a technology that brought about fundamental changes on the international stage. The development of communication technology is another; the rapid feedback it allowed fundamentally altered the nature of diplomacy. At the same time, burgeoning industrial growth has compelled nations to seek resources beyond their borders, which means technology can also increase international tensions. It is safe to say that technological evolution influences the way conflicts are triggered and how and to what extent they escalate. Given that technology continues to evolve very quickly, it is not at all clear whether the statistical patterns of war that Richardson uncovered will hold true in the future.

Are any changes apparent in the patterns of international conflict? And could we develop technologies that encourage the world to become more peaceful? In the hope of answering these questions, we set aside the statistics and traveled to Stockholm to consult an expert who has profound experience with the roots of conflict, international diplomacy, and how competition for oil, food, and water might trigger wars. Hans Blix is the former director general of the International Atomic Energy Agency (IAEA), former chief of the United Nations Monitoring, Verification and Inspection Commission that was sent to Iraq, and a doctor in international law. Discussing these matters with him, it became clear that many of the subjects explored in this book converge in any analysis of war and peace. Does this imply that some sort of shift is under way in the international order?[4]

"The use of force on the interstate level has decreased in recent decades," Hans Blix replies. "There are fewer reasons for war. There is no colonialism anymore; borders are fixed; there are almost no wars of conquest or of ideology. The international system has settled, and nobody seems eager to change it. During the cold war, there was always the threat of mutually assured destruction, which could even happen as a result of one person's mistake. That danger has faded. Of course, there might be a nuclear threat from Iran or North Korea, but that is unlikely to result in a global war. We have smaller threats now than in the time of the cold war. But they are more frequent."

There is still a risk, therefore, that nuclear weapons could push the conflict statistics in the wrong direction. History shows that it is impossible to ban nuclear weapons completely or to be absolutely certain of their whereabouts. But the fading threat of nuclear war has diminished public support for nuclear disarmament, Blix notes. "Social, legal, and psychological factors are important for the verification of international security. The IAEA offers certain safeguards, but it is only a watchdog. It barks; it doesn't bite. And even with the monitoring of the IAEA, there is never complete assurance; there is always a residue of uncertainty. Similarly, international law is a barrier. States do not like to breach legal obligations. Libya and Iraq were both violating the Nuclear Non-proliferation Treaty, but they were brought back into line. Once again, though, that doesn't bring total security."

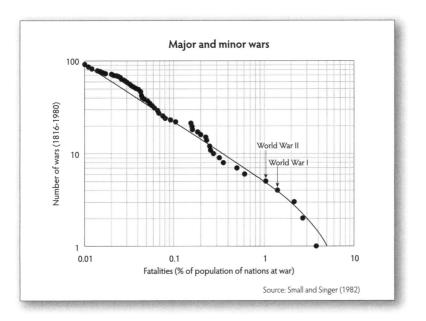

Every war seems to obey a common dynamic in terms of death toll. The frequency of major and minor conflicts displays a fixed pattern as shown in the chart, with vertically the number of wars that have at least the number of fatalities on the x-axis. This implies that there will be another large-scale war in the future. We need a clearer understanding of the underlying mechanisms if we want to avoid that. *Source*: Small, M., and Singer, J. D. (1982). *Resort to arms: International and civil wars, 1816–1980*. Thousand Oaks, Calif.: Sage Publications.

Technology may be helpful, too, Blix continues. "It can bring more transparency. Satellites are a good means because they provide eyes all over the place and have proven to be an important tool for disarmament. Environmental sampling is also a good method for the verification of nuclear activities. For example, samples of water, leaves, and cloth can be analyzed for traces of nuclear material. And there are ways of doing geological surveys to detect the underground presence of uranium. But in the field of disarmament, technology is not the final answer. It can't replace inspectors on the ground. They have a 'feel' on-site of what is happening and can draw conclusions from the attitude of recipient countries. Nations must accept the intrusive presence of inspectors." In addition, chemical weapons have an elaborate inspection regime. "For biological weapons, the formal verification options are limited," Blix says. "There are many restrictions, but there are no protocols for verification. In the biological field, one has to rely so far on ethical standards rather than on inspection because services and materials developed in these industries are multipurpose. But chemical weapons are of marginal importance, and biological weapons have the drawback of contaminating oneself."

Blix thinks it's important to understand what drives countries to develop nuclear weapons. "For Iran, it is most likely security or status. For North Korea, it's probably a way to come out of its isolation and force diplomats to speak to them. But there are other ways to do that. You should work toward integration and an increasing interdependence between states. That would decrease the risk of an international conflict. A good example is the relationship between China and Japan. There were many tensions between the two. But both are now developing trade and economic cooperation. That's the way to work toward a positive development of relations and greater international interdependence."

Hans Blix believes that the decision to pursue nuclear armament is a political one. "So the best way to avoid weapons of mass destruction is to make states feel they do not need them. Fostering mutual interdependence is therefore a hopeful approach for the future. Countries should be strongly encouraged toward this. It's not necessary to stop nuclear development in these countries altogether. You can try to shape those ambitions. Countries like Iran and North Korea can be encouraged to invest in nuclear power and research for climatic purposes. That would give them a way to use their nuclear experiments, testing facilities, and know-how for energy-saving purposes instead of weapons. That way, the interdependence of countries increases, which stabilizes the international community. Of course, there is always a residue of uncertainty. As I said, there is never a total barrier or total security. But the advantages of embedding countries in the international community are greater than the risks. It strengthens the existing international structure and takes the sting out of controversies."

Peaceful use of nuclear energy would also be a useful way to mitigate the greenhouse effect, Blix points out. "It is very valuable because it gives a huge amount of energy without carbon dioxide emissions. I am more worried about global warming than I am about weapons of mass destruction. There is an enormous difference in scale."

CHANGING THE CLIMATE

Having raised the subject of nuclear energy, climate change, and the quest for oil and other natural resources, Hans Blix points out that while the number of international conflicts might have decreased, there is one worrying exception: "We see more conflicts about resources, such as oil. But these conflicts could be played out far better in the marketplace. If prices properly reflected scarcity, it would foster technological development. There is still a lot of room for improvement. We can work on better batteries that enable new modes of transport and save oil. We can work on superconductors, which may be helpful in electricity networks. We should continue to work on nuclear fusion, which could provide a cleaner source of electricity. There are so many possibilities to provide energy without oil and without emitting carbon dioxide. We should work on it. However, we must remember that the energy problem becomes bigger as populations keep on increasing."

Blix sees nuclear power as one of the most promising possibilities. "Nuclear weapons and nuclear energy are often seen as Siamese twins, but that is incorrect. Countries can have nuclear power without nuclear weapons and the other way around. Even if you have nuclear energy, there is still a technological barrier to building weapons. It is not easy to make a nuclear bomb. The vast majority of countries have great difficulty in doing so. Especially enrichment plants are difficult to build and operate, but you do need enriched fuel in order to run a light water reactor. To generate electricity, you only need to enrich uranium to 5 percent. But there is a worry that if you can do that, you can also enrich to the some 90 percent that is needed for nuclear weapons. Therefore, the number of enrichment plants in the world should be limited. That's not easy because there is no prohibition on having them. Very few countries actually have enrichment facilities at the moment. If there were more of them, there would be more nuclear material and a greater risk of the technology being used to make weapons. To limit the number of plants, you could build international facilities. States could then be guaranteed a supply of fuel even if they didn't have enrichment plants themselves. That's the fuel bank concept. Countries would contribute nuclear fuel of various kinds and put it in depots to be distributed by the fuel bank." A country can't use low-enriched uranium for weapons if it doesn't have access to an enrichment plant. That means it could, in principle, be distributed to countries like Iran, too. But that raises certain difficult questions, Blix admits: "Should the fuel bank honor requests for fuel from countries that don't comply with nonproliferation standards? We have to think these questions through before we can start a fuel bank."

Another option is to use breeder reactors—a possibility we considered in chapter 2.3. "Fast breeder technology is available already," Hans Blix confirms. "It delivers eighty times more energy from uranium than we otherwise get. That way, our supply would last for many, many centuries, so we don't have to worry that uranium is a finite resource. The problem, however, is that a breeder reactor requires plutonium as an initial charge." Plutonium is dangerous stuff. It's extremely poisonous, and you can use it to produce weapons. "We don't want plutonium to spread," Blix stresses. "But there are other technologies as well. You can turn to thorium, for instance. It can't be enriched to make weapons, so there is no problem with weapon proliferation. There are also fewer problems with nuclear waste. And there is three times more thorium in the world than there is uranium."

EQUAL CHANCES

The ever-tighter web of mutual dependence has already led to shifts in military conflict statistics. Nobody expects France and Germany to fight a war with one another again; they now teach the same version of their history in their respective schools. The German chancellor has likewise been known to represent the French president when the latter is unable to attend an international meeting. The two countries have an integrated defense industry, and their electricity grids are increasingly intertwined, too. However, just as wars between states are occurring less frequently, other sources of violence are increasing in prominence. "Weapons of mass destruction aren't the most lethal weapons we have," Hans Blix confirms. "We have to realize that far more people die of small caliber weapons. That puts nuclear issues in perspective. The massacres in Rwanda were committed with machetes. The number of civil wars is increasing. They are fought because of inequality issues, rebellions, and minorities. States with great poverty have big problems within their borders. That's also a breeding ground for terrorism, of course. But I am not worried so much about large-scale attacks from terrorists. I know that the threat of nuclear terrorism is perceived to be high, and it is indeed real. Terrorists are difficult to deter, and they will resort to any methods. But nuclear terrorism has been overhyped in the media and politics. Creating daily anxiety is attractive for the media, but I don't think this is the greatest threat in the world. If a terrorist group were building nuclear weapons, it would probably be possible to detect that such activity was going on. And we can still do a lot more to prevent the smuggling of nuclear materials."

Blix is more concerned about smaller-scale violence. "There you see more victims. Political parties should take a rational responsibility and react accordingly. The way to eradicate it is to improve social order so that people do not feel humiliated. It's about decreasing inequality."

"People should not feel the need to seek support for their actions in religion," Hans Blix concludes. "We should work to fulfil the basic necessities of humankind—the need for water, for instance. Interesting progress has been made in breeding saline-tolerant plants so that you can use seawater for agriculture. And we should work on better ways to preserve food. There is still a significant need to work on infectious diseases. Even the common cold has many victims—especially when you're poor. And we should work toward decreasing population growth so that we can share the earth's finite resources in a fairer way. That should not be done through crude force, as it was in China, but by the education of women and by giving them health and status."

Part 6
VISION

6.0

AGENDA

We have some serious work to do. Far too many people lead miserable lives because they lack the most basic necessities to deal with hunger, thirst, shelter, disease, or disability. In addition, the prosperity currently enjoyed by many of us may not be taken for granted in the future. The experts in this book have identified a range of breakthroughs that are urgently required if we are to improve the fate of humanity in the decades ahead and look to the future with greater confidence. There will be some hard choices, and some lines of research will probably need to be pursued at the expense of others. Industry should change and adopt new strategies. And we as a society should accept and foster that change. The evolution of technology, industry, and society is a complex process full of feedback mechanisms and surprises. It's vital that we understand the most promising ways to facilitate the necessary changes of direction.

THE COMPLEXITY OF TECHNOLOGY

The technologies proposed in this book aren't straightforward; otherwise, they would have been identified much sooner. The days when you could produce a brilliant invention in your garden shed have largely gone. Anyone wishing to improve the current state of technology needs a solid pedigree and will need to labor long and hard with a group of dedicated colleagues, in many cases relying on extremely expensive equipment. Breakthroughs demand the stamina, laborious testing, and inspiration of countless scientists and engineers. Hundreds of thousands of design hours can go into a new microchip, car, or power-generation technique. Developing new technology is a complex process.

That complexity is exemplified by the development of the laser. Einstein predicted the principle of stimulated emission on which lasers are based long before World War II.[1] But it was many more decades before working lasers were created and longer still before they were put to practical use. Once we had them, however, we found we could use them in new scientific instruments that opened up fresh areas of research. This led in turn to improvements in our understanding of how particular atoms emit light, allowing the development of the improved lasers that are now mass-produced for DVD players and many other applications. There was no directed effort from Einstein to solid-state lasers; the evolutionary path twisted and turned. A new scientific discovery creates new insights, which then clear the way toward the next unexpected breakthrough. Even failures contribute to progress. During the many attempts to create small solid-state lasers, it proved extremely difficult to give them the right optical properties. When the first solid-state lasers failed to function properly, a completely unexpected sidetrack of research came up. The light emitted by these seemingly faulty devices laid the foundations for the light-emitting diode (LED).

Progress is a form of evolution in which pure science and practical applications are mutually reinforcing. It is a strongly nonlinear process that is difficult to direct. That's partly why we have limited ourselves in this book to breakthroughs with contours that are already visible rather than fantasizing about inventions that lie behind the horizon. Technological innovation has also been a process in which individuals have played a pivotal role, even if most of us are unfamiliar with their names. Dutchman Lou Ottens, for example, conceived the compact disc in the full knowledge that the lasers available at the time weren't capable of reading them properly. But the belief of people like him prompted the research that eventually made CDs possible. Visionaries like Ottens might not enjoy the high profile of a Steve Jobs, but they are no less influential.

Technology development has often been driven by outside stimuli. There are times when the pressure intensifies sharply, resulting in sudden leaps. German chemistry shifted up a gear around 1900 when the supply of saltpeter risked being cut off. This key ingredient of dynamite was primarily imported from Chile, where it was extracted from the thick layers of guano that accumulate on the country's cliff faces. Saltpeter featured prominently in the rivalry among Germany, Britain, and France—Europe's three most powerful nations at the time—and Germany risked losing out because of its poor sea access.

The stakes were high because a shortage of saltpeter could impact supplies of munitions. The Germans launched a major research effort to find a way of manufacturing the necessary nitrates artificially. A process to synthesize ammonia was quickly developed, allowing for the production of nitrogen compounds. The process was then scaled up and, just before the outbreak of World War I, Germany opened the world's first nitrate plant. The project was crowned with three Nobel Prizes—one each for Fritz Haber, Wilhelm Ostwald, and Carl Bosch, now considered the founders of modern physical chemistry.

This kind of sudden leap in the development of technology isn't made possible only by the expenditure of a considerable amount of money. The urgent need to solve problems also seems to unleash human creativity and motivate leaders to support their efforts. The more pressing the dilemma, the more resolutely we seek a solution and the wider the support that is provided. Another example occurred during World War II when hundreds of scientists were brought together in the Manhattan Project, which rapidly resulted in the building of the first atomic bomb and the first computers. The same conflict brought us synthetic rubber, major advances in radar and aerospace technology, the mass production of penicillin, and much more.

Outsiders may thus promote research when there is a pressing need, which isn't necessarily created by military conflict. Industry sponsors new technology to achieve a comparative advantage, and research can also be supported by idealistic motives, such as the humanitarian desire to provide basic requisites like clean water, food, and shelter or to help us deal with the diminishing of our planet's scarce resources. New developments hold out the possibility of new materials, safer and less wasteful production methods, and alternative sources of energy. We need to influence outside stimuli like these, given that they are crucial drivers of technological innovation. It is important to realize that they don't always lead to their goal as directly as the German empire's need to identify an alternative to saltpeter. When the first working lasers were produced in the 1960s, they were dismissed in some quarters as a solution looking for a problem. And heavy investment in agricultural technology hasn't always produced the desired results. The outcome of all the effort being expended is often far from clear; technological development is complex and hard to predict. Anyone wishing to bring about new advances needs to be aware of this and should take advantage of new insights from complexity science to increase their likelihood of success. We need to pursue more flexible solutions so that technology can serve us more effectively in a fast-changing environment. And we must also come to grips with complexity itself.

We can increase flexibility by making our technology more adaptive. A simple radio set can only pick up FM stations. But its chip can be designed to receive other bands, too, allowing the use of the same circuit for more advanced radios. Sophisticated chips with millions of components are generally cheaper to produce if they can be used for a variety of applications. It makes sense for them to be programmable, as was described in chapter 3.3. That way, they're more flexible in operation, manufacturers can easily configure them for specific applications, and they might even be able to adjust automatically to their particular use. The basic design should, at the very least, allow a much wider set of applications. Flexibility also implies the need to develop technology that requires fewer inputs and produces fewer environmental emissions—technology that can operate more independently of its surroundings, that is more energy efficient, more resistant to external disruption, and which generates less waste. Examples include the chips at the heart of a computer network, the sensors in a chemical plant, the switching stations in a power network, the catalysts in a reactor, and the electronics onboard an aircraft.

Distributed—as opposed to centralized—approaches increase flexibility, too. New process technology may allow the construction of smaller chemical plants that can produce closer to the user. We may no longer have to hide huge and polluting plants away on isolated industrial estates. In the past, we saw the dismantling of large computer centers in favor of the introduction of personal computers. Distributed energy generation is likewise gaining support in the area of electrical power networks.

Complexity-aware technology should allow us to respond faster, too. It is important in complex processes to keep our finger on the pulse, so we can respond quickly if things threaten to go wrong. Thus, it's not just a question of the big picture and long-range forecasts but also of remaining alert to small changes. We need faster, low-energy sensors so that we can measure on a much shorter timescale and with much greater accuracy than is currently possible. That's important in many technological fields: New data-processing circuitry will operate at much higher speeds; new generations of microchip will require greater precision; and we can only track climate change properly if we can monitor Earth's different systems more accurately.

Designers will also need to take into account technological flaws. As we approach the physical limits of miniaturization, it is inevitable that the microchips and nanostructures we create will begin to manifest imperfections. New techniques are needed so that fault-tolerant design can be widely applied. If faults occur in equipment, chips, or structures, they mustn't be fatal. The software that controls it all needs to cope with imperfections. Similarly, we have to accept that our knowledge of a situation is imperfect. We can't organize our logistics down to the last detail; we have to allow for glitches instead. Self-healing strategies are being pursued in telecommunications and power supply networks, for instance. If a link in the network is down, the system should be able to find an alternative route to bypass the problem. Our bodies set a good example in that regard. A broken bone can heal when the pieces are kept in place for a while; a blocked blood vessel rarely proves fatal if the process developed gradually, and other vessels can generally take over the required supply.

It's not always possible to increase flexibility. We need to improve our understanding of how technology is embedded in a larger context. One way to do that is to collect lots of data, which is the philosophy behind the robotized experiments performed by chemists. When chemistry becomes too complex for outcomes to be predicted—when seeking a new catalyst, for instance—thousands of experiments can be performed with random variations in their conditions. Powerful computers are similarly used to perform pattern recognition among random samples, and Craig Venter applies much the same approach (chapter 4.2) to as many DNA fragments as possible.

Complex technology often requires the combination of different insights. That certainly applies to the communication, energy, and transport networks we've addressed in this book. Local chemical plants clearly require a network of suppliers and distributors. If we want to take our technology further, we need to focus on the different levels—from individual building blocks to complex systems. There will be an increasing need to combine expertise from different fields. To achieve breakthroughs in terms of flexible optical chips, you need a thorough knowledge of electronics, photonics, and materials. To improve the design of telecommunication networks, you have to combine insights in optical and radio communication. And we will also need a greater understanding of the physics of materials as well as the mathematics and computer simulation techniques that can handle complex problems in all these areas.

THE COMPLEXITY OF INDUSTRY

The evolution of industry displays intriguing parallels with the gradual change of ecosystems. Both have evolved into ever greater complexity. Gone are the days when factories could produce everything they needed for themselves, as in the early days of industrialization. Those early days are comparable with an ecosystem on newly formed land—ground reclaimed from the sea, for instance, recently created dunes, or an island resulting from a volcanic eruption. The first species to establish themselves in areas like these are the ones that produce lots of seeds. These pioneers survive in a harsh environment by multiplying rapidly and spreading to new locations, thereby securing the continued existence of their species if a particular habitat is depleted. Locusts behave in a similar way: A swarm eats voraciously, quickly reproduces, and then moves on when all the food has gone. When new areas are first colonized, nature develops aggressively. But as the ecosystem continues to develop, other types of species establish themselves. Brambles—slow-growing bushes with numerous roots and branches—are a good example. Plants like this can withstand seasonal influences and survive the winter or extended periods of drought. More species then begin to appear that depend on one another; the waste products of one, for instance, provide nutrients for another. A complex biotope eventually grows up, such as those we find in an oak–beech wood or a tropical rain forest. These ecosystems can readily absorb shocks because of the wealth of species that comprise them. If there is a temporary period of wetness, the species that benefit from that will proliferate, rapidly using up the excess moisture. A self-regulating, sustainable symbiosis is thus achieved.

You can compare this succession of ecosystems with the way our industries arose and have developed. Many industries displayed locust-like behavior, especially in the first decades of the Industrial Revolution. Just as locusts care little about the fields they strip bare, factories sucked up raw materials and spewed out the residues without any thought to the depletion of natural resources or the burden they were imposing on the environment. Nowadays, there are many places where you no longer find such naked exploitation of the planet. In the course of the past century, industry set down roots, grew branches, and began to deploy more sustainable techniques. Scarce natural resources were replaced by manufactured chemical components. Industrial processes became more precise and hence more economical in their use of materials. A good example of this development is sodium carbonate—an important ingredient in the manufacture of soap. Until the end of the eighteenth century, soda was made by burning seaweed, making the product rare and expensive. The French government considered this to be a serious problem, and so in 1775, it announced that a prize would be awarded to anyone who could come up with a viable alternative. French chemist Nicolas Leblanc developed an inorganic process using sulphuric acid and salt as raw materials. It proved an immediate success. But the large volumes of chlorine and sulphur created in the process generated a new set of problems. Around the middle of the nineteenth century, Belgian Ernest Solvay chanced upon a better process while attempting to remove ammonia from coal gas. The Solvay method uses salt, ammonia, carbon dioxide, and water as raw materials, with the result that sulphur is no longer given off. Chlorine is still a problem, but nowadays, it can be extracted by electrolysis and reused. Provided everything goes according to plan, therefore, chlorine is no longer released into the environment. As the sodium carbonate industry developed over the course of the last two centuries, the processes have become much more benign and much more complex.

As industry has taken root in many places, it has evolved from pure colonization toward a symbiotic relationship with its environment. Like nature, industry can use waste materials as the input for other products, enabling it to interact more economically with its surroundings. Production is no longer a matter of sucking up raw materials and spitting out products. Industry has become a complex biotope in which factories depend on one another. Products come into existence through a long "food chain" of suppliers connected by intricate transport lines.[2] This complexity makes it hard to achieve fundamental changes. It would automatically entail a long chain of adjustments. The industrial biotope precisely reflects nature in that respect: An ecosystem can survive small changes. When a given species becomes extinct, for instance, the impact on the biotope is generally limited. Nature adapts and other species occupy the slot that has opened up. However, there are some species that play a more pivotal role in maintaining the overall balance. Their removal can precipitate the ecosystem's fragmentation and trigger a transition to a completely different one. Viewed in these terms, nature conservation is less about securing the status quo and more about preserving its dynamic ability to adapt to changes.

The supply networks in industry resemble the food networks in an ecosystem. Both display numerous short localized connections and a few long-range links that guarantee the overall balance. Network theory shows that systems of this kind arise from the pursuit of ever greater efficiency, resulting in progressively tighter interdependence. The long-range links mean that these networks are very efficient, too; there is no better way to move goods, materials, and energy streams around. That brings financial benefits and also ensures that systems of this type are very robust. The industrial network can adapt quickly to local changes just as nature adapts to disappearing species. But the downside is inertia, as was described in the first part of this book (chapter 0.2). Efficient networks are resistant to change: New methods of generating electricity struggle because large-scale, existing power plants are so firmly embedded in industrial networks. Similarly, the network of agricultural supplies and food production plants makes it hard to switch to alternative crops or products in response to fluctuations in the climate. And the Dutch built an economy powered by windmills, which was so successful that they almost missed out on the Industrial Revolution powered by the steam engine.

When a fundamental change eventually breaks through, it leads to a complete overhaul of the industrial ecosystem. Economists have long noted that industry has gone through a series of revolutions, sometimes referred to as "Kondratiev" or "Schumpeter waves," although in the context of complexity science, we would prefer to use the term "transitions." The Industrial Revolution began with James Watt's steam engine (1775); the railway revolution was spurred by intercity rail links (1830); the age of heavy industry is exemplified by the open-hearth furnace for steel production (1871); mass transport was heralded in by the first mass-produced Ford (1908); and the information revolutions were unleashed by the computer and microelectronics and by the Internet (1950s and 1960s). Industrial upheavals like this are recurrent and inevitably triggered an economic heyday—phenomena that are familiar from complex systems. The transitions in self-organizing critical systems are equally repetitive.[3]

Even though we're not prepared for major changes in our industry, we know that they're lurking around the corner. We might witness rapid shifts in climate zones, for instance, oil price fluctuations, or problems in the supply of raw materials. New technology could trigger new upheavals. Biomolecular processes, for instance, have the potential to overturn existing production structures, as exemplified by the synthesis of DNA for the first time in 2008. Industry is bound to be exposed to further shocks of that kind. As with the conservation of nature, protecting the future of our industry is not about securing the status quo but fostering the dynamics needed to adapt to changes as they arise. Industrial strategies must become more flexible. This, as we have noted, has often been at the price of temporary increases in inefficiency. But that's about to change. Miniaturization and mass-produced electronics make it possible to scale down without sacrificing efficiency. Plenty of examples in this book illustrate the point, including smart electricity grids (chapter 3.1), the chemical plant on a chip (chapter 2.5), and new approaches to logistics (chapter 3.6) and food production (chapter 1.2). Industry should embrace this approach to become more flexible and less susceptible to shocks.

THE COMPLEXITY OF SOCIETY

Technology is often just part of the solution. Social will and personal motivation can be every bit as important. The phenomena of complexity are clearly at play in the field of society. Revolutions, grassroots movements, and politicians who know exactly how to exploit a given moment. Society as a whole has characteristics that you can't accurately trace back to the behavior of individuals. New technologies may become a sudden hype, but perceived gaps in the market are often only a marketing illusion. Effective innovations can sometimes be left on the shelf for years. Advanced cryptographic security strategies, for instance, can go unused; imminent threats are often not fully perceived. There is plenty of technology capable of slowing down climate change, but it simply isn't used.

Utopian authors have always dreamed about the smooth adoption of new technologies for the benefit of society. Thomas More's

Utopia[4]—and virtually every similar vision that followed—is characterized by the total control of every aspect of society. Strong rulers enforce the introduction of new technologies to the benefit of all. In Francis Bacon's *New Atlantis*,[5] scientific research is used to conquer nature and for the betterment of society, creating "generosity and enlightenment, dignity and splendour, piety and public spirit." Yet this technocracy is by no means a pleasant place to be; its ideology of purity tolerates no dissent and turns scientists into a separate caste. And there isn't any genuine social interaction—a factor shared by all utopias.[6] Most people would flee such societies if they could. Pervasive centralized control and the suppression of interaction are certainly one way to deal with the complexity of society. But it's hardly a desirable way and would ultimately prove unstable anyway. Even the most brutal dictators are certain to overlook something sooner or later, possibly triggering their downfall. History has shown this to be true in actual dictatorships with a technologically inspired ideology, such as existed in the Soviet Union.

We certainly don't believe that our society is heading for a stable utopian end point. Managing a complex community of people is a continuous process of evolution in which rapid feedback and evidence-based decisions are as important as in any other complexity issue. This is true of every form of government. It is fascinating to see how quickly interactions via the Internet have given rise to new models of governance and a new kind of "benevolent dictator." Young Finn Linus Torvalds—chief developer of the Linux computer operating system—for instance, governs a multitude of voluntary programmers with a far from self-evident form of authority.[7] Rather than setting far-reaching goals, he devotes himself to problems as they arise, guided by constant feedback and peer review. The community of programmers debates each issue passionately, but the benevolent dictator never takes sides until the heat has gone out of the debate and evidence for the respective arguments can be properly considered. Constant and rapid feedback is needed to make this philosophy work, which is why Torvalds frequently releases new editions of Linux, so that the impact of any given decision can be observed. Unexpected side effects can then be dealt with at an early stage. These principles organize the community in such a way that only the best concepts win. There is no marketing budget or propaganda machine to conceal poor technical decisions or mediocrity.

In many countries, it is the government that facilitates the direction of technological research. Fundamental research in particular can be a very long-term process, whereby the government is often the only institution capable of supporting the extended journey from laboratory to practical implementation. As we have seen, this is frequently an iterative process that needs to be organized in stages and for which the necessary capital has to be made available. In the U.S. system, the government tends to sponsor research with a clear mission. A great deal is financed with military goals in mind, but an equally clear mission also underpins the research performed by NASA, the Institutes of Health, and the Department of Energy. There is political pressure, therefore, to work toward solutions. There is also a risk, however, that the ideological nature of the sponsorship will cause certain solutions and topics to be ignored. In Europe, decisions regarding scientific research are often taken at the member state level, resulting in fragmentation and slowing down the process of developing a unified approach. What's more, decisions regarding the research sponsored by the European Commission are not entirely based on evidence but are also prompted by the desire to give each member state a "fair turn." However, moves in the right direction are made. It is our task as scientists to encourage governments to base their decisions on more rational arguments, just as that hardheaded crowd of programmers directs the decisions that Linus Torvalds takes regarding the development of Linux. We are also responsible for identifying the side effects of new technology, which can often be a double-edged sword. The discovery of atomic power as a new source of energy, for instance, turned into a nightmare for the people of Nagasaki and Hiroshima. Those side effects can, of course, be positive, too. The synthesis of ammonia for use in munitions also led to nitrates for fertilizers, the manufacture of which fundamentally changed agriculture and was a blessing for humanity.

Emotions and outmoded opinions are another factor we need to deal with as a society. New solutions are obliged to compete with decades of shared experience, established technologies, vast, organically developed infrastructures, and production costs that have been substantially reduced by a steady flow of improvements. The more firmly a technology is established, the harder it is to displace with something entirely new. How do you compete with more than five decades of silicon technology for electronics or more than a century of gasoline engines? How do you persuade people to adopt energy-saving technologies? Our society is a complex system that is not easily shifted. The consequences of such changes are often so great that individuals will refuse to accept them until they truly have no other option.

That kind of inertia is a barrier to many new solutions. It isn't easy to ensure evidence-based decisions within government, as psychology and sociology frequently have little reliable evidence to offer. Social scientists are split into as many schools as there are football teams. That's set to change, however, as the new school of quantitative sociology is arising, as described in the last part of this book. Their results could improve our understanding of the democratic process, including the role of leaders and the media. Their research could reveal the roots of relationships and institutions and how they might be changed and enhance our understanding of herd behavior. The new approach to social sciences might also help us detect any danger of democracy sliding into dictatorship and highlight the dynamics of opinion formation, enabling us to make more balanced and rational decisions.

DEEPER INSIGHT INTO COMPLEXITY

We are only now starting to come to grips with the complexity of the challenges facing us. Our understanding of complex systems is limited because we also lack basic knowledge in many areas regarding the impact of key parameters. We continue to argue, for instance, about the climate impact of specific solar-triggered events; we are still searching for the critical factors that allow diseases to develop; and we failed to detect the adverse movement in the key parameters that unleashed the global financial crisis of 2008. Mathematics can provide us with new algorithms capable of speeding up calculations by several orders of magnitude. And we can also make the process faster by increasing our understanding of the underlying forces. We believe that our science and technology are far from mature. It is evident that there are many aspects of our planet and its inhabitants that we do not fully understand. A great deal more thus needs to be achieved if we are to succeed in protecting life on our planet.

For better understanding, we shouldn't spend time and money to look farther ahead. It will only make plain how unpredictable the future will be. That's generating confusion and fear, which is the worst possible counselor. The pressing problems of humanity call for another approach, which is reminiscent of Edgar Allan Poe's tale "A Descent into the Maelstrom." Three brothers are out fishing when their boat is caught in a terrible storm. A raging gale and furious currents turn the sea into a savage, whirling mass, dragging the boat into the center of a gigantic maelstrom, where a terrifying silence reigns. One of the brothers is profoundly affected by this. Convinced he is going to die, he regains his composure and observes with fascination everything happening around him. After a time, he notices that smaller and more cylindrical objects are being dragged down less rapidly. He then realizes that barrel-shaped objects are actually working their way upward and out of the vortex. He lashes himself to an empty water barrel and hurls himself overboard. While the wrecked fishing boat is sucked deeper and deeper, the barrel steadily rises out of the whirlpool, carrying the sole survivor to safety.

The hero of the tale saves himself by disengaging emotionally from his immediate situation. Distancing himself like this enables him to overcome his fear of death and to throw himself into the reality raging all around him. The lucid way in which he looks that reality in the face is the secret to his survival. His behavior isn't motivated by a desperate attempt to find a way out of his predicament; he's driven purely by his fascination with the maelstrom. At first, it seems as though he's frittering away what little time he has left on scientific curiosity. But his thinking doesn't go straight from problem to solution; it begins by taking a detour via pure knowledge. It's good to take a disinterested view of reality—just like that fisherman who came out of the maelstrom alive.

NOTES AND REFERENCES

0.0. OUR MISSION

1. Santen, R. A. van, Khoe, G. D., and Vermeer, B. (2006). *Zelfdenkende pillen, en andere technologie die ons leven zal veranderen.* Amsterdam: Nieuw Amsterdam.

0.1. CONCERNS

1. According to Jean Ziegler, the United Nations special rapporteur on the Right to Food for 2000 to March 2008, in his book: Ziegler, J. (2005). *L'Empire de la honte.* Paris: Fayard, p. 130.

2. His project is published in two books: Lomborg, B. ed. (2004). *Global crises, global solutions.* Cambridge, UK: Cambridge University Press; Lomborg, B. (2007). *Solutions for the world's biggest problems: Costs and benefits.* Cambridge, UK: Cambridge University Press.

3. Holdren, J. P. (2008) Science and technology for sustainable well-being. *Science*, 319, 424.

4. See the UN Web site, http://vermeer.net/caa

5. Wells, H. G. (1901). *Anticipations of the reactions of mechanical and scientific progress upon human life and thought.* Complete text, http://vermeer.net/cab

0.2. APPROACH

1. These energy forecasts are quoted in the essay "Against Forecasting" by Czech-Canadian environmental scientist Vaclav Smil. Anyone thinking about making quantitative prognoses should read it first. See Smil, V. (2003). *Energy at the crossroads: Global perspectives and uncertainties.* Cambridge, Mass., and London: MIT Press, chapter 3 and references therein.

2. See http://vermeer.net/cac

3. Lorenz, E. N. (1995). *The essence of chaos.* Seattle: University of Washington Press.

4. These earthquake statistics are known as the Gutenberg-Richter Law. See Gutenberg, B., and Richter, C. F. (1954). *Seismicity of the earth and*

associated phenomena, 2nd ed. Princeton, N.J.: Princeton University Press, pp. 17–19 ("Frequency and energy of earthquakes").

5. Per Bak himself has studied many of these subjects and has written a highly accessible introduction: Bak, P. (1997). *How nature works: The science of self-organized criticality*. Oxford, Melbourne, and Tokyo: Oxford University Press. Another interesting introduction is Newman, M. (2004). *Power laws, Pareto distributions and Zipf's law*. Arxiv preprint cond-mat/0412004; see http://vermeer.net/cad. And a popular account is Buchanan, M. (2000). *Ubiquity: The science of history…or why the world is simpler than we think*. New York: Three Rivers Press.

6. A technical overview may be found in Albert, R., and Barabási, A. L. (2002). Statistical mechanics of complex networks. *Reviews of Modern Physics*, 74(1), 47–97. A popular account is Barabási, A. L. (2003). *Linked: How everything is connected to everything else and what it means*. New York: Penguin. Also see Buchanan, M. (2002). *Nexus: Small worlds and the groundbreaking science of networks*. New York: Norton; and Watts, D. J. (2003). *Six degrees: The science of a connected age*. New York: Norton.

7. Albert, R., and Barabási, A. L. (2002) Statistical mechanics of complex networks. *Reviews of Modern Physics*, 74(1), 47–97.

8. Rockstrom, J., Steffen, W., Noone, K., Persson, A., Chapin, F. S., et al. (2009). A safe operating space for humanity. *Nature*, 461(7263), 472–475.

9. Homer-Dixon, T. (2006). *The upside of down: Catastrophe, creativity and the renewal of civilization*. Toronto: Random House of Canada Ltd.

10. The first scientist to examine this issue quantitatively was Svante Arrhenius in his 1896 paper: Arrhenius. S. (1896). On the influence of carbonic acid in the air upon the temperature of the ground. *Philosophical Magazine and Journal of Science*, 41(Series 5), 237–276.

11. Scheffer, M., Bascompte, J., Brock, W. A., Brovkin, V., Carpenter, S. R., et al. (2009). Early-warning signals for critical transitions. *Nature*, 461(3), 53–59.

1.0. VITAL NETWORKS

1. See the theme issue of the *Philosophical Transactions of the Royal Society B* (October 27, 2009). Short, R. V., and Potts, M., comp. and ed. The impact of population growth on tomorrow's world, 364 (1532).

2. The British government's Chief Scientific Adviser John Beddington, professor of applied population biology at Imperial College London, warned in 2009 that the increase in population, wealth, demand for water, and energy use would lead to an explosive situation by 2030. The figures regarding water are quoted from his statement.

3. Diamond, J. (2006). *Collapse: How societies choose to fail or succeed*. New York: Penguin.

4. The whole food chain is described in Smil, V. (2000). *Feeding the world*. Cambridge, Mass., and London: MIT Press.

1.1. WATER FOR LIFE

1. A good introduction to the field is: Sanitation and access to clean water, in Lomborg, B. (2004). *Global crises, global solutions*. Cambridge, UK: Cambridge University Press; or more recently, Rijsberman, F. (2008). Every last drop: Managing our way out of the water crisis. *Boston Review* (September/October); see http://vermeer.net/cae

2. Dai, A., Qian, T., Trenberth, K. E., and Milliman, J. D. (2009). Changes in continental freshwater discharge from 1948 to 2004. *Journal of Climate*, 22, 2773–2792.

3. Smil, V. (2000). *Feeding the world: A challenge for the twenty-first century*. Cambridge, Mass., and London: MIT Press.

4. Rijsberman, F. R. (2008). Water for food: Corruption in irrigation systems, in *Global corruption report 2008*. Cambridge, UK: Cambridge University Press.

1.2. FOOD FOR ALL

1. A good introduction to the field is Smil, V. (2000). *Feeding the world*. Cambridge, Mass., and London: MIT Press. See also Kiers, E. T., Leakey, R. R. B., Izac, A. M., Heinemann, J. A., Rosenthal, E., et al. (2008). Agriculture at a crossroads. *Science*, 320(5874), 320.

2.0.

1. Rockstrom, J., Steffen, W., Noone, K., Persson, A., Chapin, F. S., et al. (2009). A safe operating space for humanity. *Nature*, 461(7263), 472–475.

2.1. DEALING WITH OUR CLIMATE

1. Lenton, T. M., Held, H., Kriegler, E., Hall, J. W., Lucht, W., et al. (2008). Tipping elements in the earth's climate system. *Proceedings of the National Academy of Sciences*, 105(6), 1786–1793. See also Schellnhuber, H. J. (2008). Global warming: Stop worrying, start panicking? *Proceedings of the National Academy of Sciences*, 105(38), 14239.

2. An overview of the current insights may be found in a special edition of *Nature*, 458(7242), April 30, 2009, pp. 1077, 1091–1118, 1158–1166.

3. Smil, V. (2008). *Global catastrophes and trends: The next 50 years*. Cambridge, Mass., and London: MIT Press, 2008, p. 175.

4. These data and many other details can be found in the reports of the Intergovernmental Panel on Climate Change (IPCC), www.ipcc.ch. The most recent assessments are from 2007: *Climate change 2007: Synthesis*

report, Cambridge, UK, and New York: Cambridge University Press, http://vermeer.net/caz. More backgrounds can be found in the reports of the IPCC working groups that go with this report: *The physical science basis/impacts, adaptation and vulnerability/mitigation of climate change*. These reports can be downloaded at http://vermeer.net/cag

5. Schellnhuber, H. J. (2008). Global warming: Stop worrying, start panicking? *Proceedings of the National Academy of Sciences*, 105(38), 14239–14240.

2.2. IMPROVING ENERGY EFFICIENCY

1. See, for example, *BP, Statistical Review of World Energy 2009*, bp.com/statisticalreview. The historic development of oil reserves has also been analyzed by Smil, V. (2008). *Energy in nature and society*. Cambridge, Mass., and London: MIT Press.

2. Smil, V. (2003). *Energy at the crossroads: Global perspectives and uncertainties*. Cambridge, Mass., and London: MIT Press, p. 319; Stephen Pacala and Robert Socolow also showed in an influential essay in *Science* that we have the technical ability to hold carbon emissions at current levels for the next decades despite the forecast of economic and population growth. They identified fifteen available techniques that we can use to restrict emissions to current levels, including many possibilities for increasing energy efficiency. See Pacala, S., and Socolow, R. (2004). Stabilization wedges: Solving the climate problem for the next 50 years with current technologies. *Science*, 305, August 13, 2004, pp. 968–972.

3. Guzella, L., and Martin, R. (1998). Das SAVE-Motorkonzept. *Motortechnische Zeitschrift*, 59(10), 644–650.

4. Hall, C. A. S. (2004). The continuing importance of maximum power. *Ecological Modelling*, 178 (1–2), 107–113.

5. Daggett, D. L., Hendricks, R. C., Walther, R., and Corporan, E. (2007). *Alternate fuels for use in commercial aircraft*. Boeing Corporation, http://vermeer.net/cah

2.3. SEARCHING FOR NEW ENERGY

1. Niele, F. (2005). *Energy: Engine of evolution*. Amsterdam: Elsevier Science.

2. Hall, C. A. S. (2004). The continuing importance of maximum power. *Ecological Modelling*, 178(1–2), 107–113.

3. Homer-Dixon, T. (2006). *The upside of down: Catastrophe, creativity and the renewal of civilization*. Toronto: Random House of Canada Ltd.

4. This is the global means of solar radiation absorbed by Earth's surface, subtracting the energy absorbed by the atmosphere or reflected back into

space. Maxima in excess of 250 M/m2 are possible in subtropical deserts. See Smil, V. (2003). *Energy at the crossroads: Global perspectives and uncertainties*. Cambridge, Mass., and London: MIT Press.

5. See Desertec's white paper, http://vermeer.net/caj

6. The precise fraction is 16/27—the "Betz limit."

7. A state-of-the art windmill averages 1.3 W per square meter of ground area, whereas photovoltaic cells generate 3 W/m2; see Smil, V. (2008). *Energy in nature and society*. Cambridge, Mass., and London: MIT Press.

8. INL (2006). *The future of geothermal energy: Impact of enhanced geothermal systems (egs) on the United States in the 21st century*. Final Report to the U.S. Department of Energy Geothermal Technologies Program. (report number INL/EXT-06-11746). Also see Tester, J. W., Anderson, B., Batchelor, A.S., Blackwell, D.D., Pippo, di R., et al. (2007). Impact of enhanced geothermal systems on U.S. energy supply in the twenty-first century. *Philosophical Transactions of the Royal Society A*, 365(1853), 1057–1094.

9. Bell, A. T., Gates, B. C., and Ray, D. (2007). *Basic research needs: Catalysis for energy*. Report from the U.S. Department of Energy Basic Energy Sciences Workshop.

10. Klerke, A., Christensen, C. H., Nørskov, J. K., and Vegge, T. (2008). Ammonia for hydrogen storage: Challenges and opportunities. *Journal of Materials Chemistry*, 18(20), 2304–2310.

11. Loges, B., Boddien, A., Junge, H., and Beller, M. (2008). Controlled generation of hydrogen from formic acid amine adducts at room temperature and application in H2/O2 fuel cells. *Angewandte Chemie International Edition*, 47, 3962–3965.

2.4. SUSTAINABLE MATERIALS

1. Data from Ton Peijs.

2.5. CLEAN FACTORIES

1. Jensen, K. (2001). Microreaction engineering—Is small better? *Chemical Engineering Science*, 56, 293–303.

2. El-Ali, J., Sorger, P. K., and Jensen, K. F. (2006). Cells on chips. *Nature*, 442, 403–411.

3.0. OUR ASSISTANTS

1. McLuhan, M. (1962). *The Gutenberg galaxy: The making of typographic man*. London: Routledge & Kegan Paul.

2. Drexler, K. E. (1986). *Engines of creation*. Garden City, N.Y.: Anchor/Doubleday.

3. Joy, B. (2000). Why the future doesn't need us: Our most powerful 21st-century technologies—robotics, genetic engineering, and nanotech—are threatening to make humans an endangered species. *Wired,* 8(4), 238–264.

4. Kurzweil, R. (2005). *The singularity is near: When humans transcend biology.* New York: Viking.

3.1. SMARTER ELECTRONICS

1. Mollick, E. (2006). Establishing Moore's law. *IEEE Annals of the History of Computing,* 28(3), 62–75.

2. For the latest roadmap, see *International technology roadmap for semiconductors,* http://www.itrs.net

3. Ross, P. E. (2008). Why CPU frequency stalled. *IEEE Spectrum,* 45(4), 72.

4. The former CTO of Philips Semiconductors drew a lot of attention when he observed that a linear increase in functionality requires an exponential increase in system complexity. Claasen, T. (1998). The logarithmic law of usefulness. *Semiconductor International,* 21(8), 175–186.

3.2. MORE COMMUNICATION

1. *Network World* (January 22, 1990).

2. This is now a well-studied phenomenon. See, for example, Barabási, A. L., and Albert, R. (1999). Emergence of scaling in random networks. *Science,* 286(5439), 509–512; Albert, R., Jeong, H., and Barabási, A. L. (2000). Error and attack tolerance of complex networks. *Nature,* 406, 378–382; Albert, R., Jeong, H., and Barabási, A. L. (1999). The diameter of the World Wide Web. *Nature,* 401, 130–131; Barabási, A. L. (2001). The physics of the Web. *Physics World,* 14(7), 33–38.

3. Dorren, H.J., Calabretta, N., and Raz, O. (2008). Scaling all-optical packet routers: How much buffering is required? *Journal of Optical Networking,* 7(11), 936–946.

3.3. REACHING EVERYONE

1. Haykin, S. (2005). Cognitive radio: Brain-empowered wireless communications. *IEEE Journal on Selected Areas in Communications,* 23(2), 201–220.

2. Hoekstra, J. M., Van Gent, R. N. H. W., and Ruigrok, R. C. J. (2002). Designing for safety: The "free flight" air traffic management concept. *Reliability Engineering and System Safety,* 75(2), 215–232.

3.4. CRYPTOGRAPHY

1. Records are listed by Paul Zimmermann; see http://vermeer.net/cak

2. These mechanisms are described in Anderson, R., and Moore, T. (2006). The economics of information security. *Science*, 314(5799), 610–613; Anderson, R. (2001). *Why information security is hard: An economic perspective*. ACSAC, Proceedings of the 17th annual computer security applications conference, p. 358.

3. For a good introduction to the topic, see Mermin, N. D. (2007). *Quantum computer science: An introduction*. Cambridge, UK: Cambridge University Press.

4. Singla, P., and Richardson, M. (2008). Yes, there is a correlation: From social networks to personal behaviour on the Web. in *Proceedings of the seventeenth International Conference on the World Wide Web (WWW'08)*, pp. 655–664.

3.5. MANAGING FAILURES

1. Joint study by ESI International and Independent Project Analysis, press release (July 31, 2008).

2. Ibid.

3. Bowen, J. P., and Hinchey, M. G. (2006). Ten commandments of formal methods...ten years later. *Computer*, 39(1), 40–48.

4. Mainzer, K. (2007). *Thinking in complexity: The complex dynamics of matter*, 5th ed. Berlin, Heidelberg, New York: Springer Verlag.

5. Hopfield, J. J. (1982). Neural networks and physical systems with emergent collective computational abilities. *Proceedings of the National Academy of Sciences*, 79(8), 2554–2558.

3.6. ROBUST LOGISTICS

No notes.

3.7. ADVANCED MACHINES

1. See http://www.worldrobotics.org

4.0. THE NURSERY OF LIFE

No notes.

4.1. THE TRANSPARENT BODY

1. *Deaths, percent of total deaths, and death rates for the 15 leading causes of death: United States and each state, 1999–2005*. U.S. Department of Health and Human Services, 2008.

2. Zaidi, H. (2006). Recent developments and future trends in nuclear medicine instrumentation. *Medical Physics*, 16(1), 5–17.

3. Zaidi, H., and Prasad, R. (2009). Advances in multimodality molecular imaging. *Journal of Medical Physics*, 34(3), 122–128.

4. Result are not yet conclusive; cf. Krupinski, E. A., and Jiang, Y. (2008). Anniversary paper: Evaluation of medical imaging systems, *Medical Physics*, 35(2), 645–659.

4.2. PERSONAL MEDICINE

1. Despite their intense rivalry, Craig Venter and Francis Collins announced the mapping of the human genome jointly, accompanied by then U.S. president Bill Clinton.

2. Much of the network analysis we outline here is based on Goh, K. I., Cusick, M. E., Valle, D., Childs, B., Vidal, M., and Barabási, A. L. (2007). The human disease network. *Proceedings of the National Academy of Sciences*, 104(21), 8685–8690.

3. Venter talks about the disease known as familial adenomatous polyposis (FAP). It is triggered by a defect in the APC gene.

4. Kim, E., Goren, A., and Ast, G. (2008). Alternative splicing: Current perspectives. *BioEssays*, 30(1), 38–47.

5. The key point here is that proteins are knocked out at random, which means there is a high probability that the small number of proteins that play a pivotal role in the cell will be spared. See Nacher, J. C., and Akutsu, T. (2007). Recent progress on the analysis of power-law features in complex cellular networks. *Cell Biochemistry and Biophysics*, 49(1) 37–47; Jeong, H., Mason, S., Barabási, A. L., and Oltvai, Z. N. (2001). Lethality and centrality of protein networks. *Nature*, 411(11), pp. 41–42.

6. Levy, S., Sutton G., Ng, P. C., Feuk L., Halpern A. L., et al. (2007) The diploid genome sequence of an individual human. *PLoS Biology*, 5(10), e254, 2113–2114.

4.3. PREPARING FOR PANDEMICS

1. See Bird flu deal hangs in the balance. *New Scientist* (November 17, 2007).

2. Johnson, N., and Mueller, J. (2002). Updating the accounts: Global mortality of the 1918–1920 "Spanish" influenza pandemic. *Bulletin of the History of Medicine*, 76(1), 105–115. For a dramatized account, see Barry, J. M. (2004). *The great influenza: The story of the deadliest pandemic in history*. New York: Penguin.

3. Debora MacKenzie, Will a pandemic bring down civilization? *New Scientist* (April 5, 2008).

4. See http://vermeer.net/can

5. The Macroepidemiology of Influenza Vaccination (MIV) Study Group. (2005). The macroepidemiology of influenza vaccination in 56 countries, 1997–2003. *Vaccine*, 23, 5133–5143.

6. See http://vermeer.net/cap

7. FAOSTAT 2007, http://faostat.fao.org

8. According to a study by Oliver Wyman, a consulting firm, together with the WHO. Press release (February 23, 2009), http://vermeer.net/caq

9. Swine flu: How experts are preparing their families. *New Scientist* (August 12, 2009), issue 2721.

10. More than 95 percent of the world's seasonal influenza vaccine is produced in Australia, Canada, France, Germany, Italy, Japan, the Netherlands, the United Kingdom, and the United States. Smaller production facilities are located in Hungary, New Zealand, Romania, and Russia. Fedson, personal communication.

11. Data from BD Medical Surgical Systems, 2007, cited in McKenna, M. (2007). *The pandemic vaccine puzzle*. CIDRAP. www.cidrap.umn.edu

12. Fedson, D. S. (2008). Confronting an influenza pandemic with inexpensive generic agents: Can it be done? *Lancet Infectious Diseases*, 8(9), 571–576; Fedson, D. S. (2009). Confronting the next influenza pandemic with anti-inflammatory and immunomodulatory agents: Why they are needed and how they might work. *Influenza and Other Respiratory Viruses*, 3(4), 129–142; Fedson, D. S. (2009). Meeting the challenge of influenza pandemic preparedness in developing countries. *Emerging Infectious Diseases*, 15(3), 365–371.

13. Fedson, D. S. (2009). Confronting the next influenza pandemic with anti-inflammatory and immunomodulatory agents: Why they are needed and how they might work. *Influenza and Other Respiratory Viruses*, 3(4), 129–142.

4.4. QUALITY OF LIFE

1. *World population prospects: The 2008 revision*. New York: United Nations.

2. Vijg, J., and Campisi, J. (2008). Puzzles, promises and a cure for ageing. *Nature*, 454(7208), 1065–1071.

3. Wiegel, F. W., and Perelson, A. S. (2004). Some scaling principles for the immune system. *Immunology and Cell Biology*, 82, 127–131.

5.0. ENGINEERING SOCIETY

1. An introduction to the field is Lazer, D., Pentland, A., Adamic, L., Aral, S., Barabási, A. L., et al. (2009). Computational social science. *Science*, 323(5915), 721–723. See also Epstein J. M., and Axtell, R. L. (1996). *Growing*

artificial societies: Social science from the bottom up. Washington: The Brookings Institution.

5.1. ESSENTIAL EDUCATION

1. Literacy in those countries stands at 24.0 and 23.6 percent, respectively. *United Nations Development Programme Report 2007–2008*, p. 226.

2. Meltzoff, A. N., Kuhl, P. K., Movellan, J., and Sejnowski, T. J. (2009). Foundations for a new science of learning. *Science*, 325(5938), 284–288.

5.2. MAINTAINING IDENTITY

1. Recently, for example, Greenfield, S. (2004). *Tomorrow's people: How 21st century technology is changing the way we think and feel*. London: Penguin; Greenfield, S. (2008). *I.D.: The quest for identity in the 21st century*. London: Sceptre.

5.3. PROSPECTS OF CITIES

1. The first study seems to be that of Auerbach, F. (1913). Das Gesetz der Bevölkerungskonzentration. *Petermanns Geographische Mitteilungen*, 59(13), 73–76. It has since been studied intensively by Zipf, G. K. (1941). National unity and disunity: The nation as a bio-social organism. Akron, Oh.: Principia Press.

2. Gabaix, X. (1999). Zipf's law for cities: An explanation. *Quarterly Journal of Economics*, 114(3), 739–768; Pumain, D. (2002). *Scaling laws and urban systems*, http://vermeer.net/cas; Semboloni, F. (2008). Hierarchy, cities size distribution and Zipf's law. *The European Physical Journal B*, 63(3), 295–302.

3. May, R. M. (1988). How many species are there on Earth? *Science*, 241(4872), 1441–1449. There is, however, one significant statistical outlier: *Homo sapiens* is 10,000 times more abundant than should be the case based on the size curve. Hern, W. M. (1990). Why are there so many of us? Description and diagnosis of a planetary ecopathological process. *Population and Environment*, 12(1), 9–39.

4. The amount of energy used by a single living cell in an animal decreases with the one-quarter power of its total weight. Kleiber, M. (1932). Body size and metabolism. *Hilgardia*, 6, 315–351. Further thoughts on this issue can be found in Smil, V. (2000). Laying down the law. *Nature*, 403(6770), 597.

5. This was pointed out in West, G. B., Brown, J. H., and Enquist, B. J. (1997). A general model for the origin of allometric scaling laws in biology. *Science*, 276(5309), 122–126. The precise explanation remains rather controversial among biologists. In our view, however, the paper by Geoffrey West et al. offers a plausible explanation.

6. Bornstein, M. H., and Bornstein, H. G. (1976). The pace of life. *Nature*, 259(19), 557–559; Bornstein, M. H. (1979). The pace of life: Revisited. *International Journal of Psychology*, 14(1), 83–90.

7. Bettencourt, L., Lobo, J., Helbing, D., Kühnert, C., and West, G. B. (2007). Growth, innovation, scaling, and the pace of life in cities. *Proceedings of the National Academy of Sciences*, 104(17), 7301.

8. Florida, R. (2002). *The rise of the creative class: And how it's transforming work, leisure, community and everyday life.* New York: Basic Books.

9. Bettencourt, L. M. A., Lobo, J., Strumsky, D., and West, G. B. *The universality and individuality of cities: A new perspective on urban wealth, knowledge and crime.* In press.

10. Castells, M. (2000). *The rise of the network society*, 2nd ed. Oxford, UK: Blackwell Publishers.

11. Dhamdhere, A., and Dovrolis, C. (2008). Ten years in the evolution of the Internet ecosystem. *Proceedings of the eighth ACM SIGCOMM conference on Internet measurement*, 183–196.

12. Townsend, A. M. (2001). Network cities and the global structure of the Internet. *American Behavioral Scientist*, 44(10), 1697–1716.

13. Sassen, S. (2001). *The global city: New York, London, Tokyo.* 2nd edition. Princeton, N.J.: Princeton University Press.

14. Batty, M. (2008). The size, scale, and shape of cities. *Science*, 319(5864), 769–771; Batty, M. (2005). *Cities and complexity: Understanding cities with cellular automata, agent-based models, and fractals.* Cambridge, Mass., and London: MIT Press.

5.4. DISASTER SCENARIOS

1. Dilley, M., Chen, R. S., and Deichmann, U. (2005). *Natural disaster hotspots: A global risk analysis.* Washington, D.C.: World Bank Publications.

2. Ibid.

3. Tsunamis' aftermath/warning signals, but no warnings: Early data on Asian quake went unnoticed in Vienna. *International Herald Tribune* (December 29, 2004).

4. McNicol, T. Japan lays groundwork for national earthquake warning system. *Japan Media Review* (April 13, 2006).

5. Jonkman, S. N. (2007). *Loss of life estimation in flood risk assessment—Theory and applications.* PhD thesis, Delft University, the Netherlands.

6. Helbing, D., Ammoser, H., and Kühnert, C. (2006). Information flows in hierarchical networks and the capability of organizations to successfully respond to failures, crises, and disasters. *Physica A*, 363(1), 141–150; Buzna, L., Peters, K., Ammoser, H., Kühnert, C., and Helbing, D. (2007).

Efficient response to cascading disaster spreading. *Physical Review E*, 75(5), 56107–56108; Dodds, P. S., Watts, D. J., and Sabel, C. F. (2003). Information exchange and the robustness of organizational networks. *Proceedings of the National Academy of Sciences*, 100(21), 12516–12521; Helbing, D., and Kühnert, C. (2003). Assessing interaction networks with applications to catastrophe dynamics and disaster management. *Physica A: Theoretical and Statistical Physics*, 328(3–4), 584–606.

7. Berkhout, A. J. (2000). *The cyclic model of innovation*. Delft, the Netherlands: Delft University Press.

5.5. RELIABLE FINANCE

1. As David Viniar, CFO of Goldman Sachs, put it: "We were seeing things that were 25 standard deviation moves, several days in a row." *Financial Times* (August 13, 2007).

2. Milton Friedman and Eugene Fama developed this idea in the 1950s for the financial markets, and a great deal of investment know-how is based on it. The theory doesn't require everyone to behave rationally: From time to time, someone makes an incorrect calculation, causing prices to fluctuate a little. But this is a chance process generating random and small-scale financial market movements. Since there will always be shrewd investors who know how to take advantage of a price that is temporarily too high or too low, standard theory dictates that they will drive the price back toward its correct equilibrium.

3. Bouchaud elaborates on this idea in an essay he wrote at the height of the 2008 financial crisis: Bouchaud, J.-P. (2008). Economics need a scientific revolution. *Nature*, 455, 1181. See also Bouchaud, J.-P. (2009). The (unfortunate) complexity of the economy. *Physics World*, 22, 28–31; Buchanan, M. (2009). Meltdown modelling. *Nature*, 460, 680–682; Farmer, J. D., and Foley, D. (2009). *Nature*, 460, 685–686. The economy needs agent-based modelling.

4. An illuminating discussion between a classical economist and a champion of nonlinear dynamics can be found in Farmer, J. D., and Geanakoplos, J. (2008). *The virtues and vices of equilibrium and the future of financial economics*, http://vermeer.net/cau

5. Bak, P., Paczuski, M., and Shubik, M. (1997). Price variations in a stock market with many agents. *Physica A*, 246, 430–453.

6. Joulin, A., Lefevre, A., Grunberg, D., and Bouchaud, J.-P. (2008). *Stock price jumps: News and volume play a minor role*, http://vermeer.net/cav

7. Numerous hedge funds got into difficulty in 2008 because of their use of the Black-Scholes model. The equilibrium theory on which that model draws underestimates the likelihood of crashes. Ten years earlier, Black-Scholes had earned its inventors a Nobel Prize in economic sciences.

5.6. PEACE

1. Holdren, J. P. (2008). Science and technology for sustainable well-being. *Science*, 319, 424–434.

2. Piepers, I. (2006). *Dynamics and development of the international system: A complexity science perspective*, http://vermeer.net/caw

3. Herrera, G. L. (2006). *Technology and international transformation: The railroad, the atom bomb, and the politics of international change*. Albany: State University of New York Press.

4. Blix, H. (2008). *Why nuclear disarmament matters*. Boston, Mass., and London: MIT Press.

6.0. AGENDA

1. Einstein, A. (1917). Zur Quantentheorie der Strahlung. *Physikalische Zeitschrift*, 18, 121–128.

2. See Axtell, R. L. (2001). Zipf distribution of U.S. firm sizes. *Science*, 293 (5536), 1818–1820; May, R. M. (1988). How many species are there on Earth? *Science*, 241(4872) 1441–1449.

3. Devezas, T. C., and Corredine, J. T. (2002). The nonlinear dynamics of technoeconomic systems—An informational interpretation. *Technological Forecasting and Social Change*, 69, 317–358.

4. More, T. (1518). Libellus vere aureus, nec minus salutaris quam festivus, de optimo rei publicae statu deque nova insula Utopia. Basileae: apud Io. Frobenium.

5. Bacon, F. (1626). *The new Atlantis*, http://vermeer.net/cax

6. Achterhuis, H. (1998). *De erfenis van de utopie*. Amsterdam: Ambo.

7. Rivlin, G. (2003). Leader of the free world: How Linus Torvalds became benevolent dictator of Planet Linux, the biggest collaborative project in history. *Wired Magazine*, no. 11.

ACKNOWLEDGMENTS

The preparation of this book has been a journey that started in 2006 with a series of interviews on the occasion of the fiftieth anniversary of Eindhoven University of Technology in the Netherlands. Later, we broadened our scope with a new and more international series of discussions. We had long and intense conversations with about fifty experts. We are very grateful for their hospitality, their enthusiasm, and the inspiration they triggered in us. It is an honor to bring so much brainpower and vision together in one book. These experts are, of course, not responsible for the text in this book; all opinions are ours except when we explicitly quote others.

We thank Ted Alkins (Leuven, Belgium) for his language advice and clear editing. His creativity contributes much to conveying the messages of this book. We thank Nel de Vink (Essex, UK) for making the graphs. Sybe Rispens (Berlin, Germany) and Thijs Michels (Eindhoven, the Netherlands) gave valuable comments on parts of this book. Publisher Nieuw Amsterdam (Amsterdam, the Netherlands) published an earlier and more limited version of this book in Dutch and was kind enough to consent to this international edition. We thank Yulia Knol for compiling the index.

The work on this book was made possible by a grant from two top research schools in the Netherlands: NRSC-Catalysis and COBRA.

INDEX

Note: Page numbers in *italics* refer to graphs.